London Trams and Trolleybuses

Michael H.C. Baker
With additional information by Russell Fell

TOP DECK PUBLISHING

Reviving the memories of yesterday…

© Images and design: The Transport Treasury 2024. Text: Michael Baker.

ISBN 978-1-913251-32-1

First published in 2024 by Transport Treasury Publishing Ltd., 16 Highworth Close, High Wycombe, HP13 7PJ
www.ttpublishing.co.uk

Printed by Short Run Press Ltd., Exeter.

The copyright holders hereby give notice that all rights to this work are reserved.
Aside from brief passages for the purpose of review, no part of this work may be reproduced, copied by electronic or other means, or otherwise stored in any information storage and retrieval system without written permission from the Publisher. This includes the illustrations herein which shall remain the copyright of the copyright holder.

Front Cover: London Transport tram class E1 No 1352 on route 10 at the City & Southwork terminus. *(Milepost 92½)*

Frontispiece: Feltham tram 2121 on route 24 (Tooting-Embankment) heading along the Embankment on a winter's day, possibly in early 1949, shortly before 2121 was sold off to Leeds City Tramways. At Leeds it was renumbered to 558. It was finally withdrawn from service in September 1957 and burnt the following month.
LT tram route 24 ran from the Embankment to Tooting Broadway. It came off in January 1951 as part of the stage 2 tram closure programme, to be replaced by new route 104, which was operated from Clapham (CA) garage using RTLs. *(A. R. Gault)*

Rear Cover: L3 1424 is seen terminating at Parliament Hill Fields on a wet winter's day. The blind has already been set for its next return on the 615 to Moorgate. *(Nick Nicholson)*

Introduction

Most of the photographs within these pages were taken around 1950, some earlier, a few much earlier, some a little later. It was all a long time ago. Nearly all the vehicles depicted are long gone, although the preservationists have served London wonderfully well so a diverse group, a few rather illogically, although none the worse for that, survive. Much of the background has also changed vastly, although some of it hasn't, and although fashion moves at a new Elizabethan, jet powered rate, it also goes round in circles. Nevertheless one can predict the date of any picture pretty accurately, certainly within a year or so, by observing what the passers by are wearing, especially the women.

More than half the vehicles depicted are either trams or trolleybuses. The last of the latter departed from the streets of London (by which we, of course, include the suburbs and beyond) in 1962, never to return, whilst the last tram headed for incineration at Charlton, close to the ground of the football team who twice around the same time reached the FA Cup Final, winning once, ten years earlier. Trams have returned, in a somewhat hesitant manner, in the Croydon area and, had London the foresight and courage of cities such as Birmingham and Manchester, would have spread much further; they may yet.

Had not the Second World War intervened, trolleybuses would have replaced all the trams, even by 1940 its fleet had become the largest in the world. After the war for various reasons London decided that diesel buses would replace not only the remaining trams but, eventually, the trolleybuses too. Trams were seen as old fashioned, the cause of traffic congestion and both trams and trolleybuses by virtue of being tied to their electric supply, either from the conduit or overhead, 'unsightly' wire, were inflexible. Oil was still relatively cheap and, after all 'we had our own,' as it was asserted by successive governments, by virtue of the 51% share in the Anglo Iranian company which Britain had owned since 1908. Never mind that in the new, post-colonial world of the New Elizabethan era Iran would attempt to nationalise its oil in 1951, the company being renamed BP (British Petroleum) in 1954, and Britain would go to war in 1956 over the nationalisation of the Suez Canal in 1956, through which oil tankers sailed to the UK.

The motor bus fleet was in such a poor state of repair after the inevitable neglect of the war years, that London Transport's priority was to first patch up and then get rid of much of it whilst the tougher, although also neglected tram fleet, could soldier on for a few more years. The first of the long promised post-war version of the AEC RT motor bus appeared in 1947 and production of it and its Leyland equivalent, the RTL and the RTW, grew so rapidly that by 1949 London was able to put 1592 new buses in service, more than the total number of the entire fleet of any other British bus company. It was now time to get rid of the trams and this was accomplished in the summer of 1952. Many regret their demise. And anyhow electricity is on its way back.

Michael H. C. Baker,
March 2023

Brentford High Street was, thank goodness, not typical as this scene in York Street, Twickenham demonstrates. Tram No.61 has the reasonably wide street to itself, apart from one horse driven vehicle and a barrow. It was one of the original, 1899 vintage Hurst Nelson four wheelers with accommodation for 39 passengers on top and 30 below. *(Milepost 92½)*

Part I
Trams

Above. In Merton Road and about to turn into Merton High Street is E1 No.1828, some five minutes into its journey fromAbove. Wimbledon Town Hall to the Embankment. The E1, built by the London County Council, was in production between 1907 and 1930, being already outdated by the time the last took to the streets, despite which they would survive for another 22 years. We shall meet many more E1s, in their various guises, throughout this book.
(A. R. Gault)

Top left. Hammersmith Broadway, c1911. The London County Council was destined to become far and away the largest operator of trams in the London area. Here are two of its standard E1s, Nos. 989 and 1578, dating from 1907 and 1911 respectively, and two London General B type buses. LCC tracks got within a few feet of the LUT's but such was the antipathy of the two concerns that they refused to join the gap. This, of course, delighted the General and its buses reaped the rewards, sewing the seeds for the eventual complete defeat of the tram in London. *(Milepost 92½)*

Bottom left. Two E1s at the same location bound for their southern terminus by Wimbledon Town Hall and Wimbledon station. Today Tramlink terminates in the station. *(A. R. Gault)*

One does not usually associate ladies in fur coats with tram travel. Presumably this lady walking past an E1 has no intention of boarding. *(A. R. Gault)*

Opposite Top: Purley was the southernmost extremity reached by the London tram network. In this scene, full of wonderful period detail, the conductor of tram No. 393 keeps a protective eye on his alighting passengers, two women well laden with shopping and a toddler. Behind, the driver of an almost new Ford V8 Pilot saloon waits patiently, illustrating one of the inconveniences, indeed what could sometimes be hazards, of tram tracks in the middle of a busy main road, in this case what had been, until a bypass was opened in the 1930s, the main A23 London to Brighton road. Behind the camera is the crossroads where the by-pass and the old main road meet and the A22 to Eastbourne begins. The scene, apart from the disappearance of the tram tracks, looking northwards towards Croydon, has changed little. The tram is a former Croydon Corporation Hurst Nelson car of 1927/8 and was, in reality, a rather up market version of the LCC E1 with seats, 'upholstered in best quality moquette.' Owing to some very old fashioned Metropolitan Police notions, it would be more than ten years before windscreens could be fitted, tough luck on the driver. *(Milepost 92½)*

Bottom: A wet, windy day in 1950 at the Purley terminus of the 16/18 routes. Passengers are belting across the London Road to gain the shelter of E3 No.1942 which has already set its blind for the return journey to the Victoria Embankment. London Transport got rid of all the elderly and very outdated Croydon Corporation trams preceding the Corporation E1s of 1927/8, in the 1930s, replacing them with ex-LCC E1s. These in turn were replaced by a batch of E3s c1939 so that Croydon was served exclusively from then on by some of the most modern trams in the entire fleet. Which probably accounted for the fact that there were many Croydonians who saw no need to get rid of the tram in 1951 and, decades on, welcomed its revival, in the shape of the immensely successful Tramlink system of May, 2000. Heading away from the camera is an RT on route 166, which, after April, 1951 became part replacement for tram route 42. *(A. R. Gault)*

Opposite Top: Feltham No.2128 is standing outside Purley depot, 7th April, 1951, the conductor replacing the pole on to the overhead whilst the driver is approaching. Further down the road a horse drawn United Dairies milk cart is heading home to its depot, next to the tram one, whilst further on is a London bound E3. There were 100 standard Felthams, constructed in 1929/30 for the Metropolitan Electric Tramways and the London United Tramways. *(A. R. Gault)*

Opposite Bottom: To say the Feltham was generations ahead of any other tram to be found in London, or, indeed, elsewhere in the UK, is no understatement. Nowadays the tram is seen as the saviour of our cities, at least in Manchester, Birmingham and elsewhere and – of course – Croydon, to say nothing of abroad. But the revolutionary Felthams of 1929/30 came too late. The powers that be in charge of the London Passenger Transport Board had decided the tram had to go. The Feltham, typified by this one, No. 2136 outside Purley depot, almost at journey's end, was big, bold and beautiful, it was fast, it was quiet, it was comfortable, drivers and the travelling public loved it. The author went to school on a Feltham often and even today, nearly 70 years since it last graced the streets of London and the southern suburbs, to visit the National Tram Museum at Crich and board No.331, the only working surviving Feltham, does not feel like a trip back in the past but one comfortably in the present. *(A. R. Gault)*

Above: Purley Depot on the afternoon of 7th April, 1951. Photographers prepare to record rebuilt E1 No.839 which is being got ready to perform the rites as the last tram to work the Purley route. It had been hired for the occasion by the Croydon Chamber of Commerce. *(A. R. Gault)*

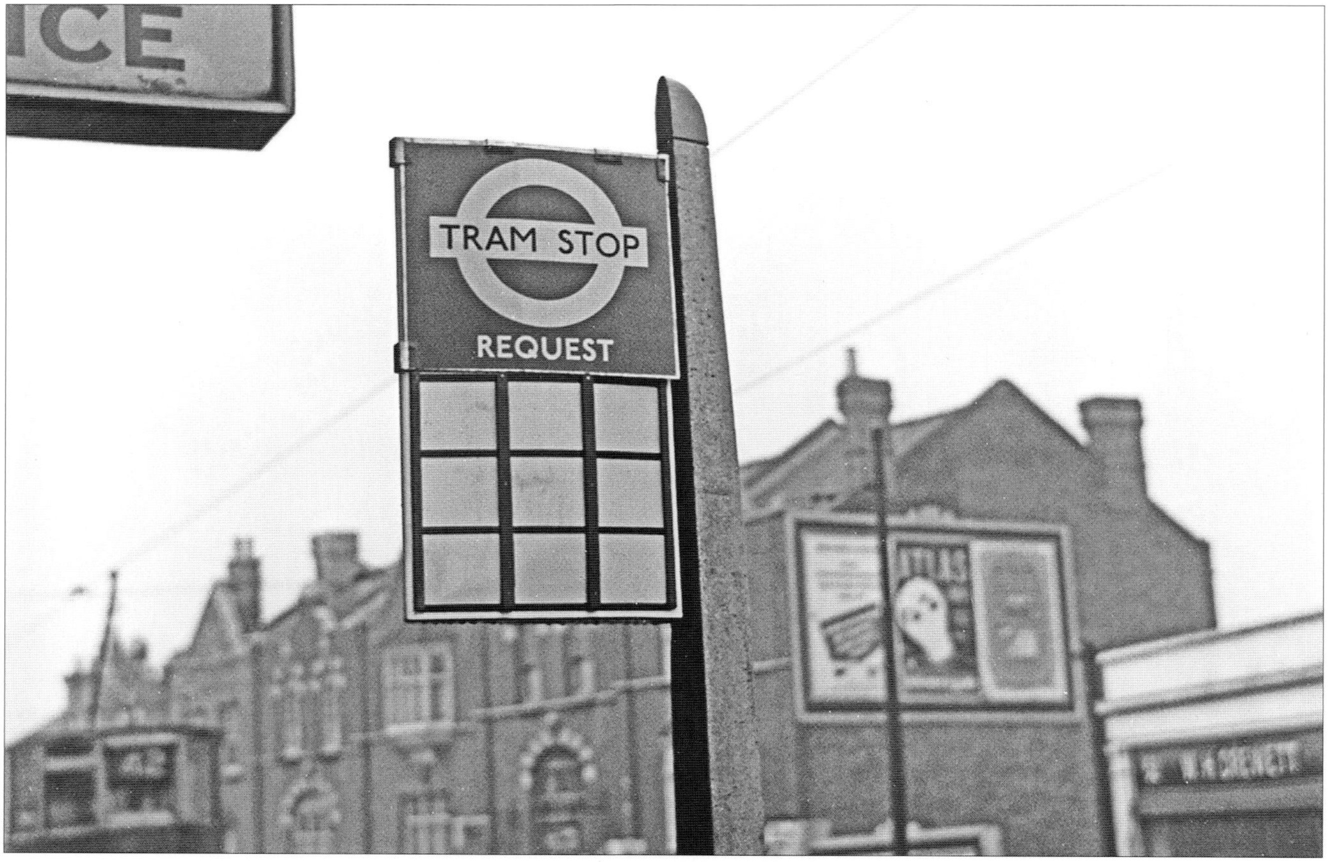
TRAM STOP
REQUEST

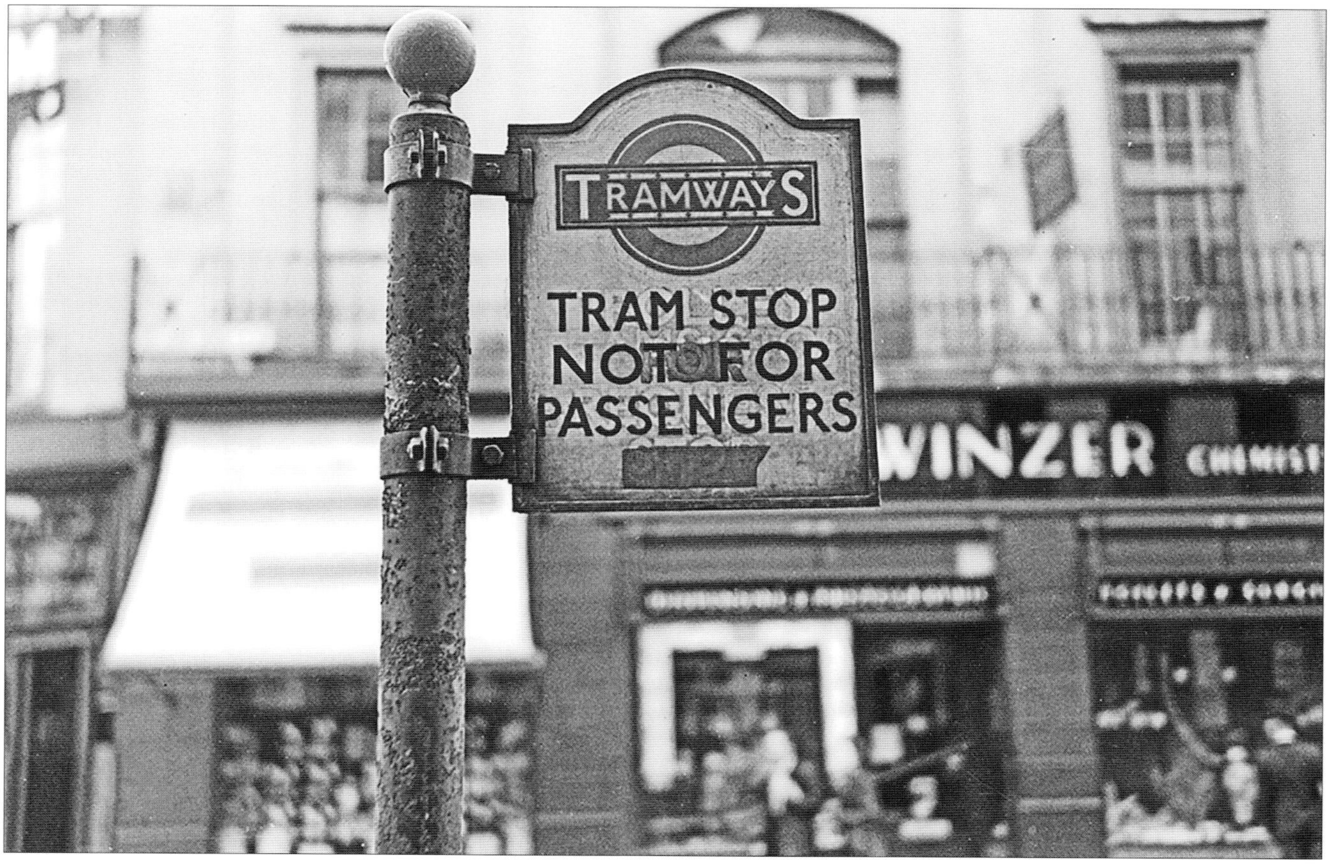
TRAMWAYS
TRAM STOP
NOT FOR
PASSENGERS
WINZER

Opposite Top: Tram stop on route 42 near Thorton Heath Library. This shows new bus stop sign with a temporary tram stop hood. After the last tram has passed on conversion night, the hood will be removed and E-plates added to show the buses stopping there. *(A. R. Gault)*

Opposite Bottom: Most tram stops were of pre-London Transport design, such as this former LCC one at Vauxhall; although, fortunately, very few were actually telling prospective passengers to clear off. *(A. R. Gault)*

Above: Dartford Urban District Council tram 10 in Dartford High Street in the summer of 1906. The line ran between Horns Cross (Greenhithe) to the east of Dartford and the boundary with Bexley via Crayford. The line opened in February 1906 and photographs of this line are relatively rare. No10 was one of a batch of 12 trams ordered from United Electric Cars of Preston. The livery was maroon and white. The whole fleet was destroyed by a fire at the depot in 1917. In this picture the tram is entering a single tram section where the high street narrows. *(Milepost 92½)*

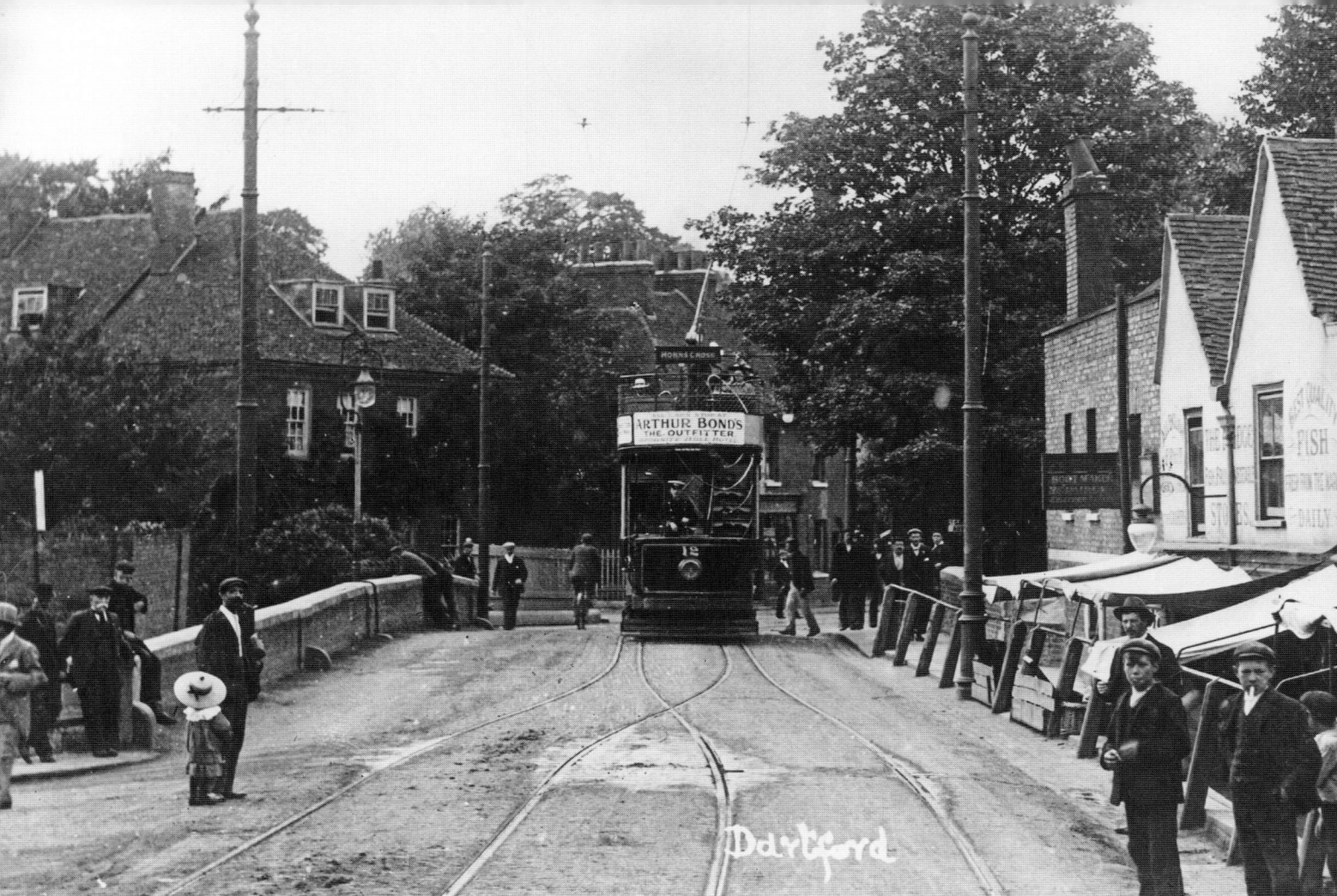

Above: Dartford tram 12 heading east towards Horns Cross (Greenhithe). The tram tracks overlap as the road crosses the bridge of the River Darenth. Once over the bridge the High Street becomes known as Overy Liberty before climbing East Hill. The unusual word 'Overy' dates from medieval times and is possibly a corruption of 'over the river'.
(Milepost 92½)

Opposite Top: The main London to Brighton Road at Norbury, c1905. The driver of Croydon Corporation car No.32, identical to No.21 in the previous picture, appears to be waiting for the ladies hurrying across the road. One can understand why the voluminous costumes of the Edwardian era gave way to something more practical when a journey by public transport became the norm for women. Kennards store, which the tram is advertising, was opened in central Croydon in 1853 and lasted until 1973. It had a sure grasp of publicity, its success largely reliant upon attracting women customers. It kept its prices as low as possible, and in order to persuade families to make a day out of a visit there was an arcade accommodating a small funfair, numerous slot machines, and donkey rides.
(Milepost 92½)

Opposite Bottom: The Norbury terminus of the LCC Embankment routes, c1912. In the distance, under the railway arch, is a four wheel Croydon Corporation car. It was not until 7th February, 1926 that the connection between the two systems was made and through running between Purley and the Victoria Embankment became possible. The trams were in competition with the electric trains, which passed over the railway bridge, in connecting the suburbs with central London. Trams took longer than trains to get there but cost less. E1 No.1027 was built in 1908 and was withdrawn just before the outbreak of the Second World War. Close numerical neighbour, No.1025, has been preserved and is part of the London Transport Collection. *(Milepost 92½)*

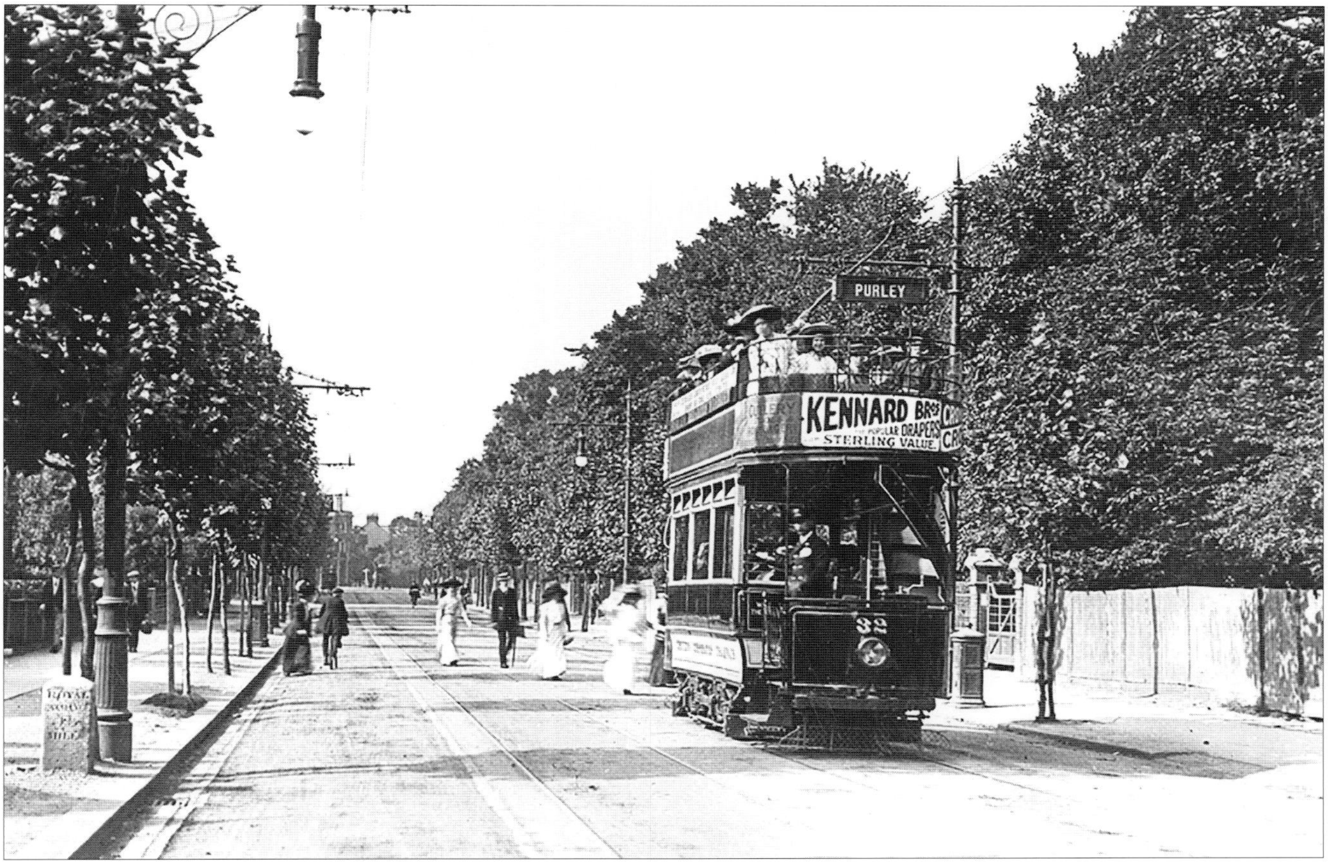
PURLEY
KENNARD BROS
STERLING VALUE
32

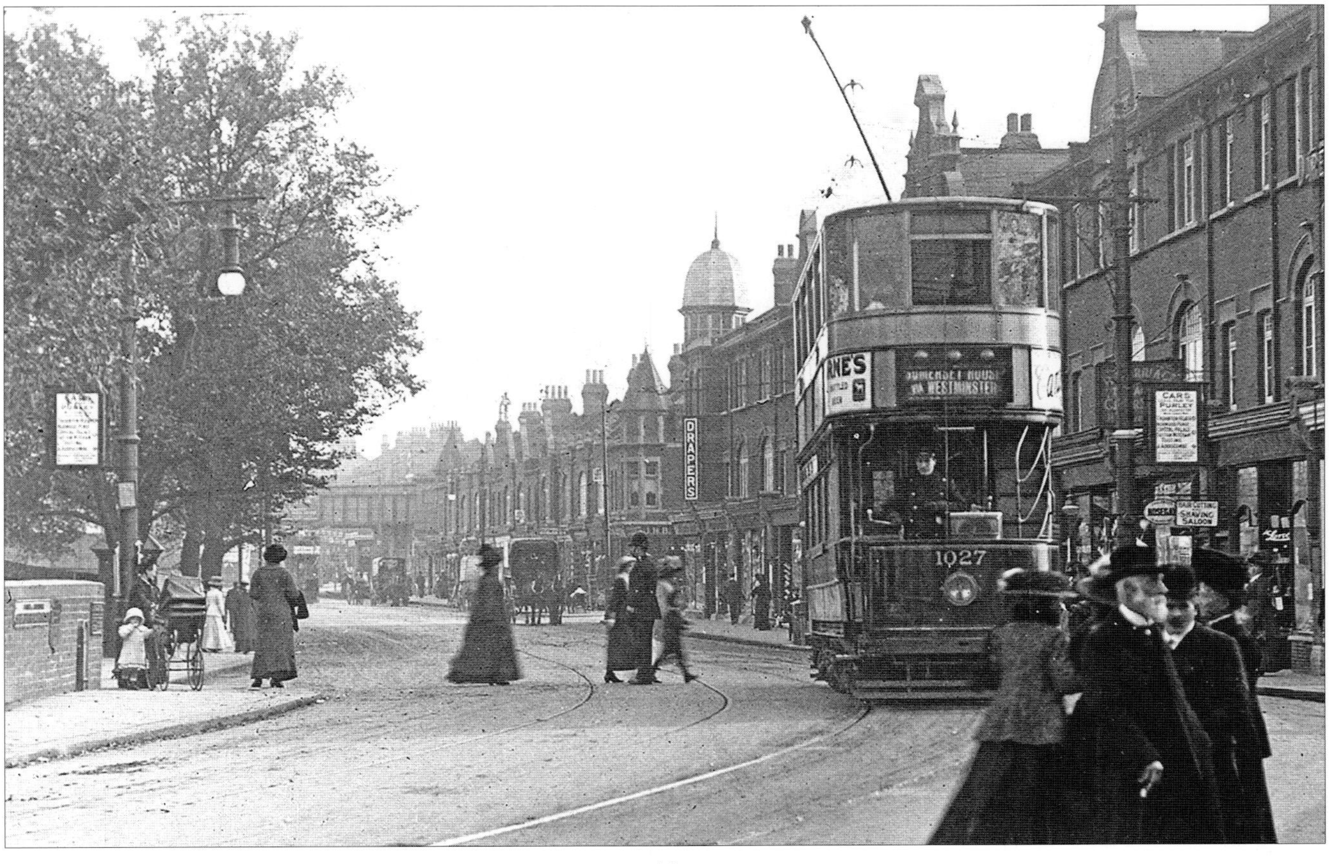
SOMERSET HOUSE
VIA WESTMINSTER
DRAPERS
1027
CARS
PURLEY
SHAVING SALOON

Above: Former Croydon Corporation tram No.378 has just arrived at the Thornton Heath terminus of route 42. Once trams carried on past the typical c1900 school building down the road to connect with the Croydon to Crystal Palace route but that had long been abandoned when this picture was taken in 1950. The 133 bus proved to be capable of absorbing the traffic. RT1175 is on the opposite side of the road working a route which, unlike the 42, still exists although it no longer serves either Thornton Heath or Croydon. *(A. R. Gault)*

Opposite Top: No.1 outside Telford Avenue Depot. This very fine tram was completed in 1932 to the design of G.F. Sinclair who had recently taken over at the LCC. Until then the LCC's policy had been to maintain all trams in a satisfactory state but not to innovate – for instance most of the fleet still had no windscreens through the 1930s, and the E1 remained in production from 1907 through to 1930. Nearly as big a leap forward as the Felthams, No.1 remained a one off and did very little work. It had air brakes and only four drivers were qualified to drive it so its appearances were few and far between. This photograph was taken on 7th April, 1951, the last day of trams to Croydon when the celebrated driver and author, Stan Collins, took it down to Purley and back and then returned it to the depot whence it was sent off to Leeds along with the Felthams. Subsequently preserved, as this is being written its restoration to its 1932 condition in the unique blue livery in which it appeared is being completed at the National Tram Museum at Crich. *(A. R. Gault)*

Opposite Bottom: No.2167. This experimental Feltham entered service with the Metropolitan Electric Tramways in October, 1929, in other words before the LCC's E3s and HR2s. It was light years ahead of anything in the LCC fleet and one can only regret that London Transport could not see that the Feltham could have transformed street transport in London. No.2167 was kept in reserve for much of the war but emerged shortly afterwards and worked regularly on Feltham duties out of Telford Avenue depot until being broken up in 1949. *(A. R. Gault)*

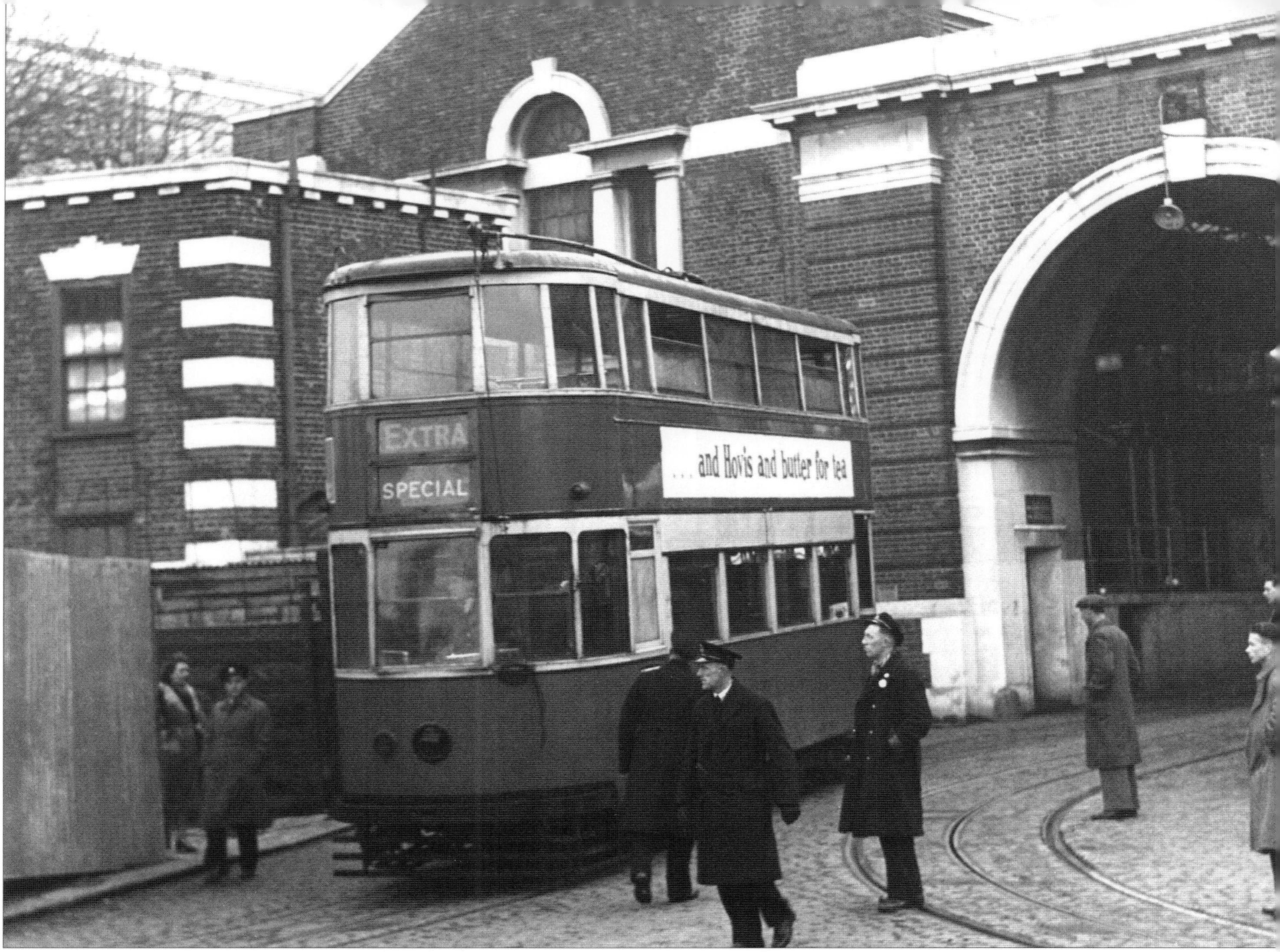
EXTRA
SPECIAL
... and Hovis and butter for tea

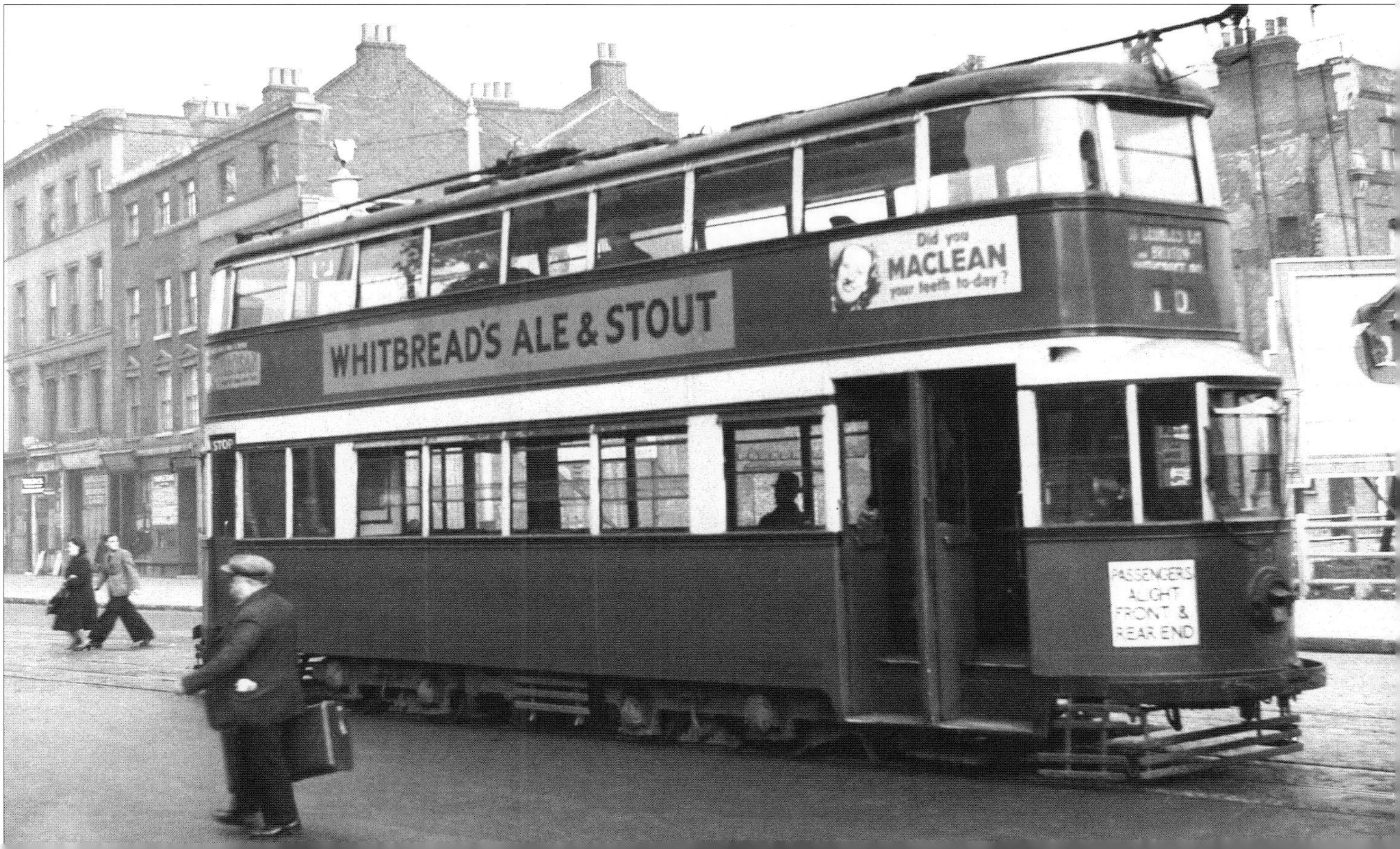
Did you
MACLEAN
your teeth to-day?
WHITBREAD'S ALE & STOUT
PASSENGERS
ALIGHT
FRONT &
REAR END

Above: New Cross Gate at the very beginning of the LCC's tram system in 1905. Amongst the various four wheel and bogie cars is a solitary horse bus. *(Milepost 92½)*

Opposite Top: E1 No.1660 passing through New Cross Gate on its way to Victoria, 6th October, 1951. New Cross was the largest London depot with accommodation for 314 cars. Like so many trams in later days No.1660 appears to have suffered dents below the windscreen. The three bulges above the indicator date back to when No.1660 was new in 1912 and were coloured lights, combinations of which were route indicators for those unable to read, of which there were a fair number in pre-1914 days. New Cross depot towards the end was not known for its high standards of maintenance. *(A. R. Gault)*

Opposite Bottom: E3 No.2000 at the Effra Road, Brixton change over during an early April shower in 1952. Several ploughs are taking a breather before being inserted back into the nether regions of trams bound for central London – even if prejudice prevented them from reaching right into the heart of it. With such a distinctive number this particular tram has made sure it has a perfectly fitted windscreen. It is being followed by a motor cycle combination, a very popular form of transport in the 1950s, and one of the pre-war designed RTs. *(A. R. Gault)*

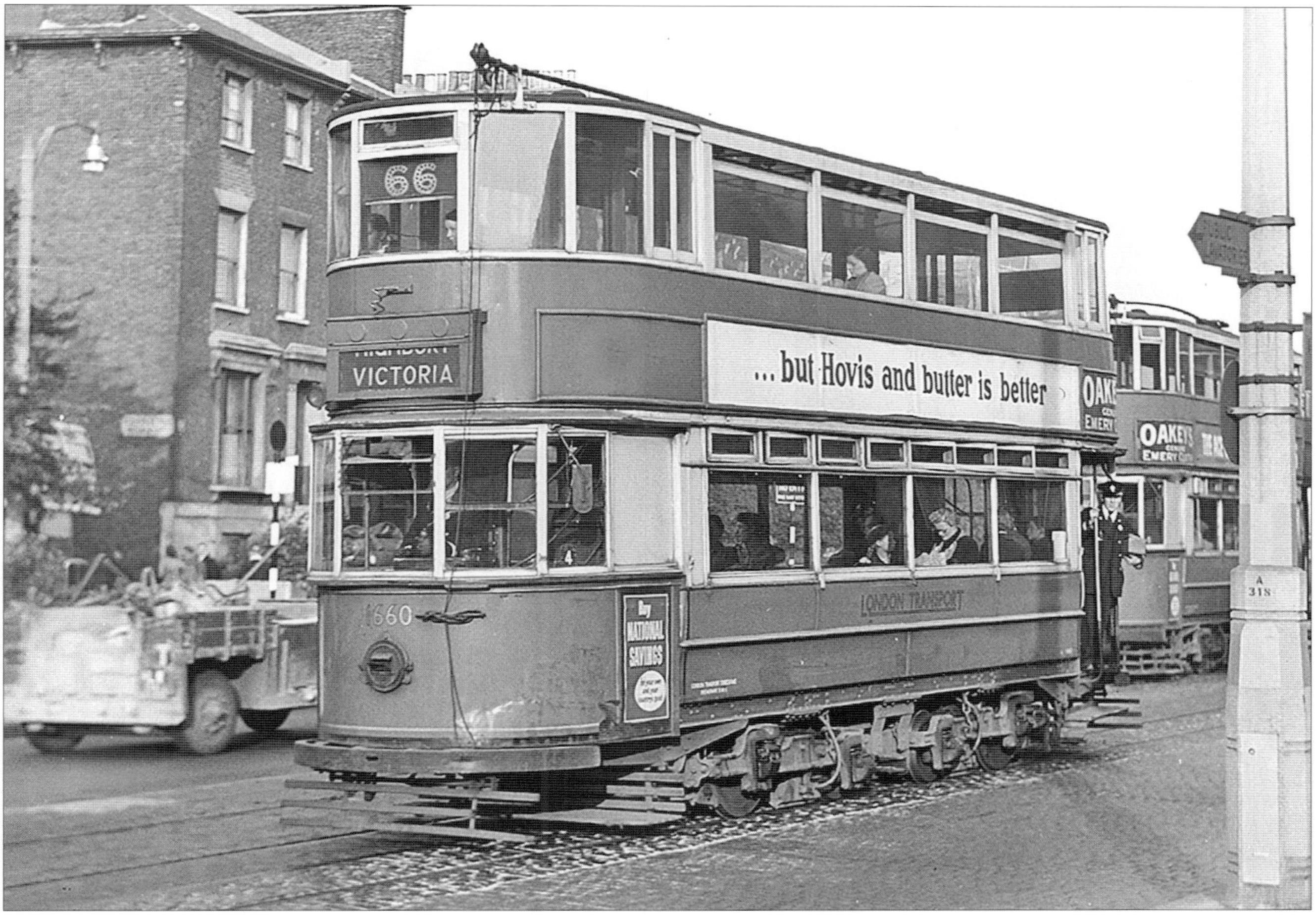
66
VICTORIA
... but Hovis and butter is better
Buy
NATIONAL SAVINGS
LONDON TRANSPORT
OAKEY'S
EMERY

33
... and Hovis and butter for tea
KINGSWAY SUBWAY
WEST NORWOOD
Esso

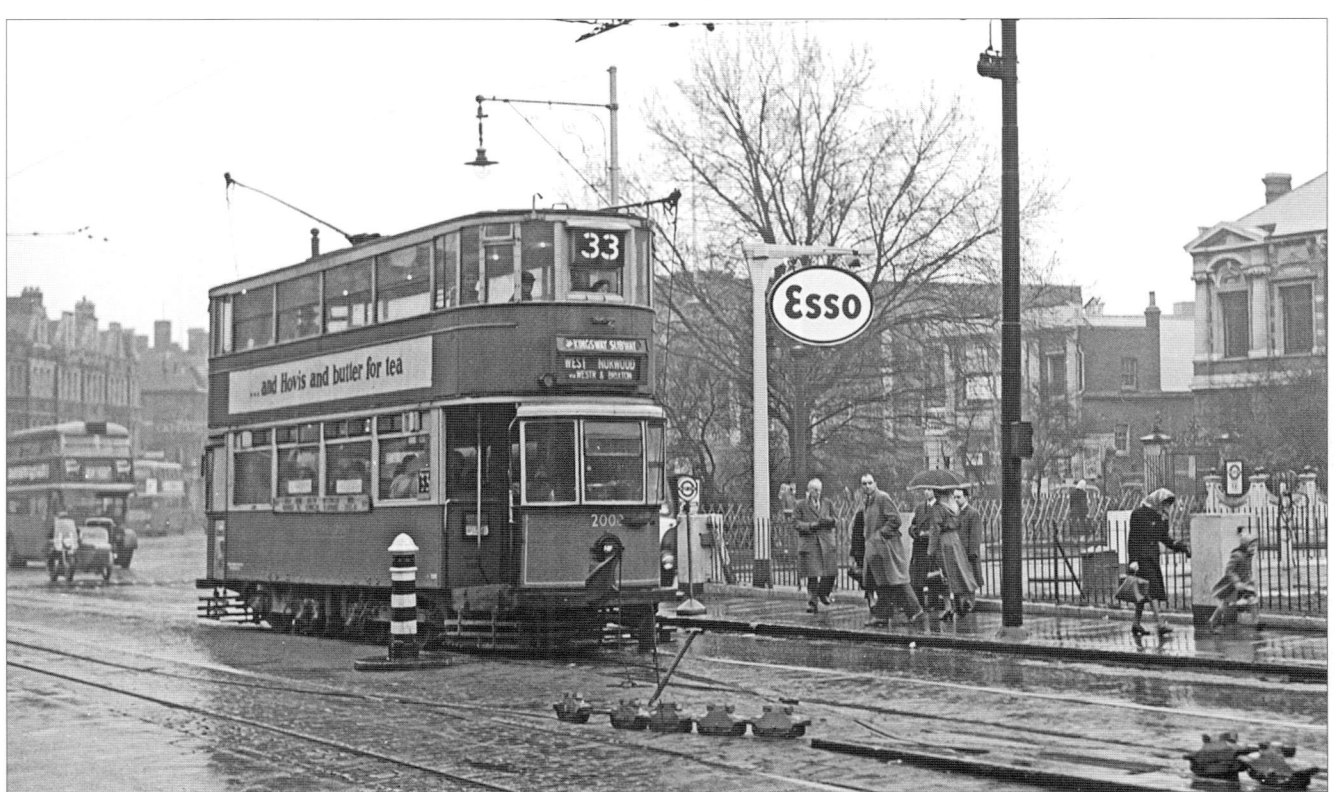

Top: The Autumn sunshine – the leaves have not yet fallen - beams down on E1 No.1494 as it heads along Peckham Road, 6th October, 1951. *(A. R. Gault)*

Left: A much graffitied No.2043, one of the former Walthamstow Corporation E1s, fitted with powerful, rather noisy motors and a bulbous, distinctively tubby windscreen, performing the last rites of route 52 on Lewisham Road, 5th January, 1952. *(A. R. Gault)*

Top: Tram E3162 passing the ornate pumping station on Green Lanes shortly after leaving its Manor House terminus. The pumping station orginally served the reservoirs which once straddled Green Lanes. It is currently used as a climbing centre. *(A. R. Gault)*

Bottom: No.015 makes a slow but steady climb up Lavender Hill away from Clapham Junction, pursued by motor buses of the D and STL variety. It is overtaking a Bradford van, which sold in some numbers in the late 1940s and early 1950s. 015 belonged to the works fleet, a sand carrier, a rebuild of an ancient LCC four wheel passenger car. Works cars retained their open platforms to the end, giving this driver, for instance, an unpleasant reminder of days gone by. *(A. R. Gault)*

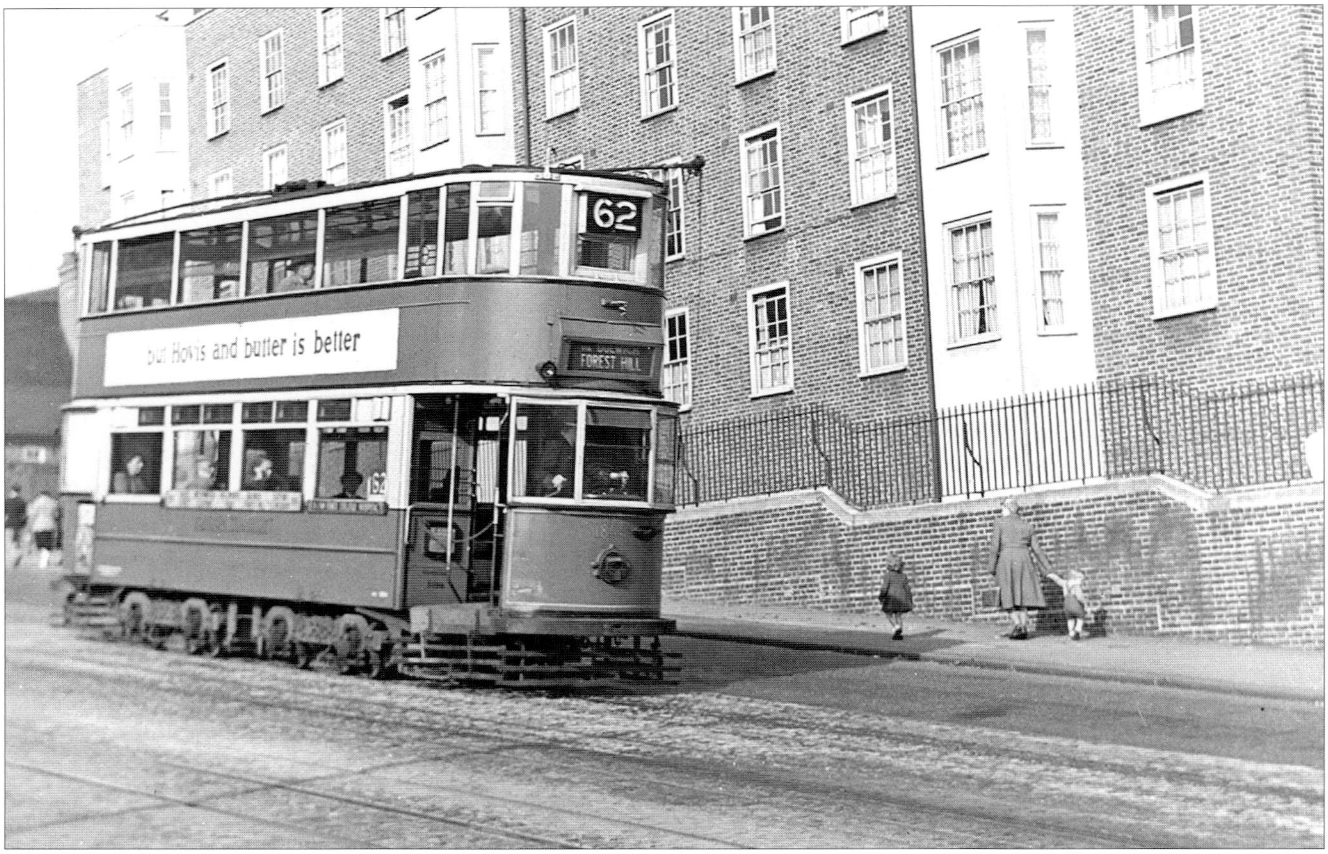
62
but Hovis and butter is better
FOREST HILL

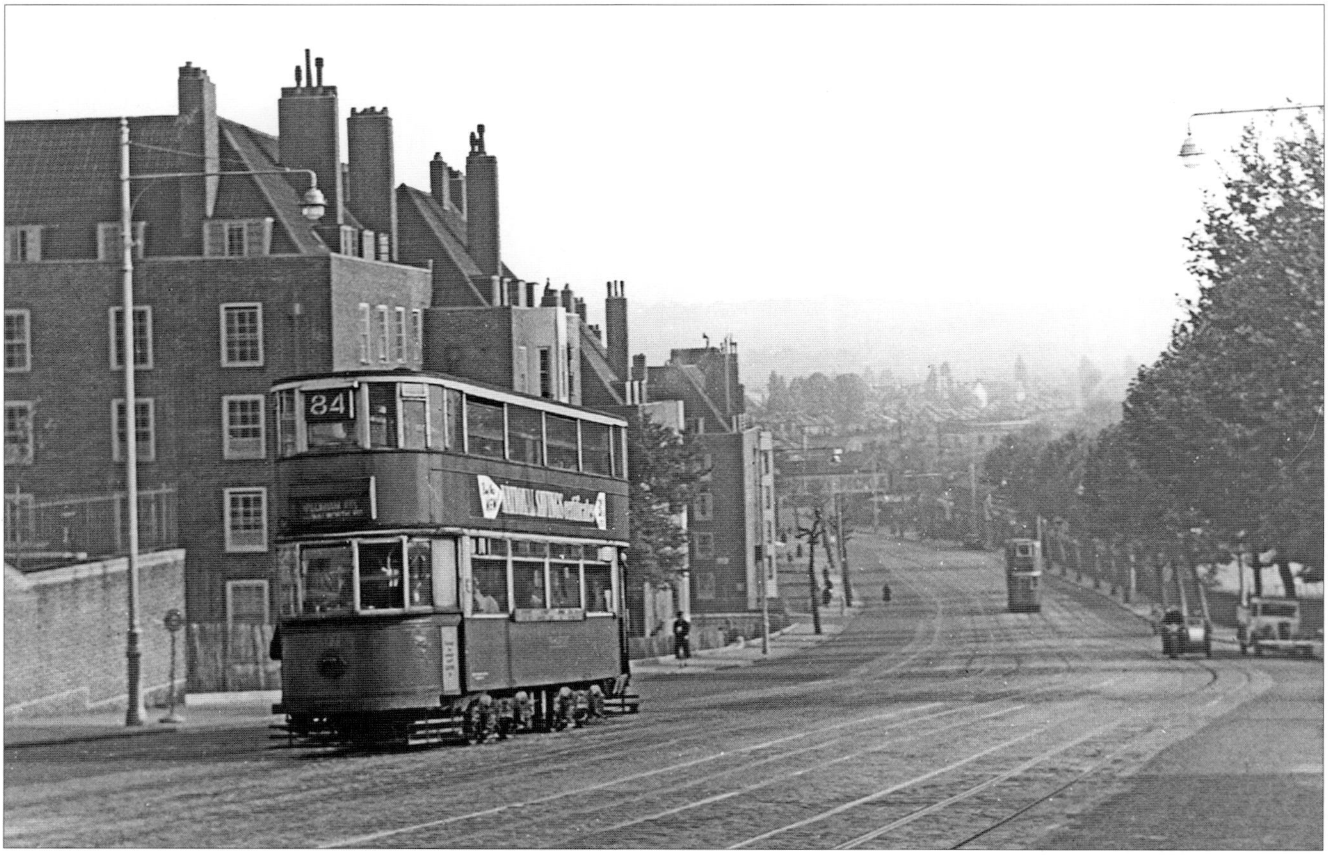
84

Opposite Top: Another 6th October, 1951 view. The driver of HR2 (Hilly Route) No.1854 keeps a tight hold of his controls as he descends Dog Kennel Hill. Many of the HR2s were without trolley poles, never operating beyond the conduit. A family is making its way up, not without some hesitation on the part of at least one member. One wonders if, later, he reflected that the London tram was an everyday feature of his earliest years. The HR2s were contemporary with the E3s with identical bodies but with four motors and bogie wheels of equal size. There were two HR1 prototypes, both destroyed in the Blitz in Camberwell Depot. *(A. R. Gault)*

Opposite Bottom: A third October, 1951 view, looking down Dog Kennel Hill past the LCC estate towards central London. HR2 No.122 is descending, in the distance is another tram whilst a motor cycle combination is making the ascent, overtaking a pre-war Austin light truck. Three HR2s were sold to Leeds in 1939 and lasted beyond the end of the London system. It is still possible to ride an HR2, No.1858 being rescued after the London system was abandoned. It spent some time on display at Chessington Zoo, quite a way off the beaten track, before being transferred very much further, but with London trolleybus companions, to Carlton Colville Museum on the East Anglian Coast.
(A. R. Gault)

Above: A busy scene in Lordship Lane, Dulwich before the First World War. A brand new LCC four wheel C class tram car No.300 of 1907 is heading up West to Victoria, the only other traffic being the horse driven cart parked in the right foreground. There are three varieties of head gear on display in front of it on what is clearly a warm, sunny day, the gentleman ambling across the road with cap, cigarette and a parcel under his arm is clearly not expecting his stroll to be disrupted by any further vehicles. The C class spent much of their careers in the hilly Dulwich area, being withdrawn in the late 1920s. *(Milepost 92½)*

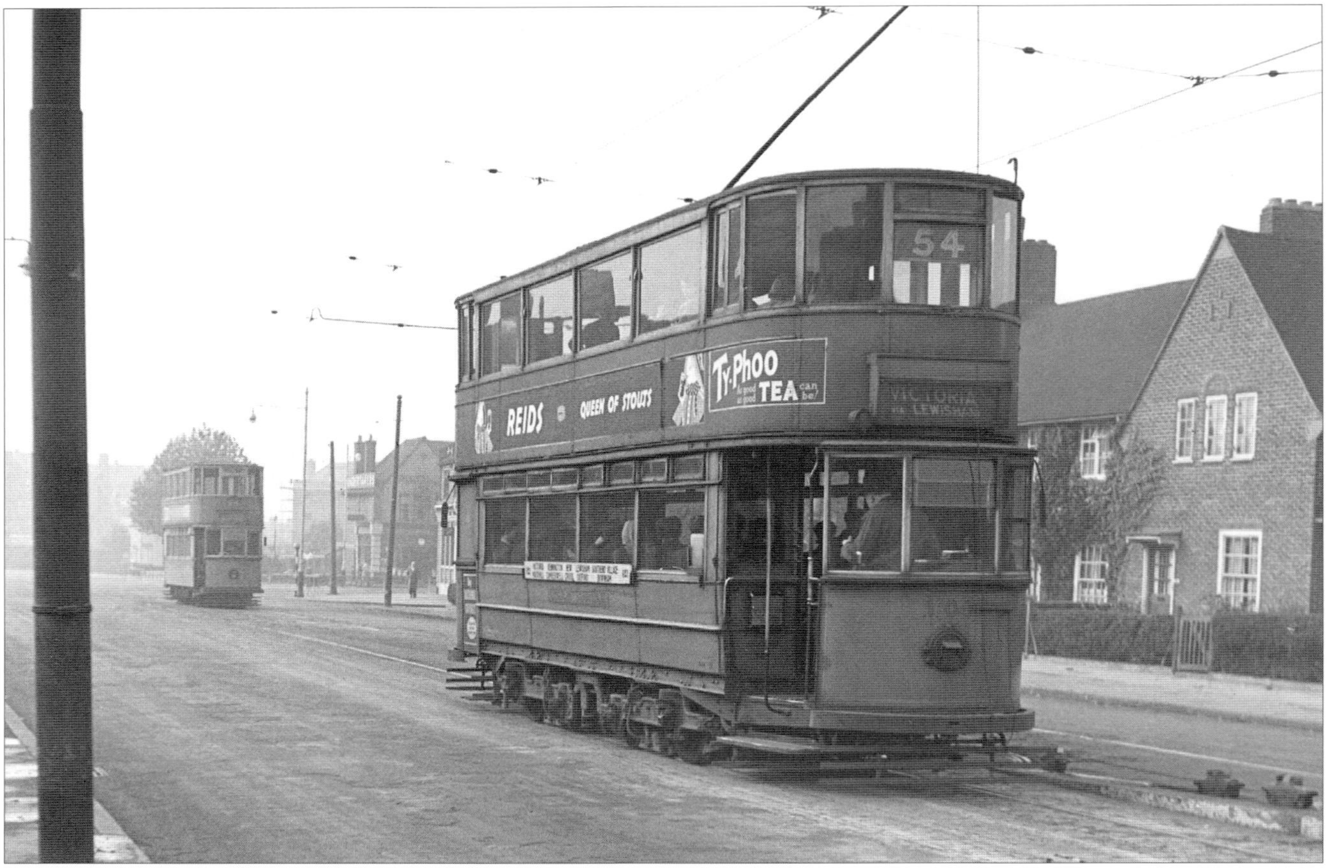
54
REIDS
QUEEN OF STOUTS
Ty-Phoo
TEA
VICTORIA

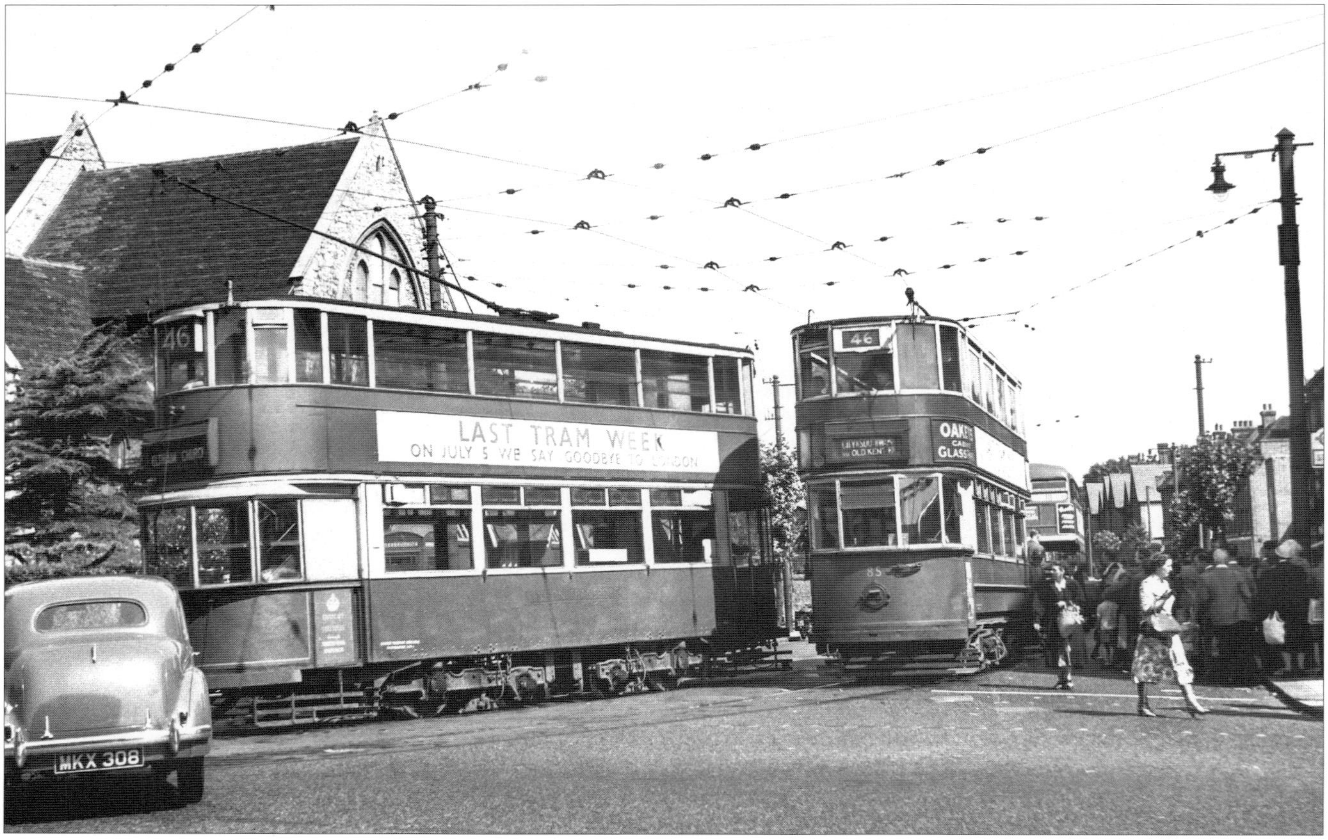
46
LAST TRAM WEEK
ON JULY 5 WE SAY GOODBYE TO LONDON
46
MKX 308

Above: A former East Ham E1 glides up the long, between the wars, development of Well Hall Road. In the distance is an E3, away on the horizon is Woolwich. The date is 16th April, 1952. It was only in the last days of the London tram that E3s became commonplace in south-east London. The citizens are going about their business on what looks like a pleasantly warm Spring day. Parked is an Austin 7 of late 1930s vintage, the type of car on which a large proportion of the aspiring middle classes learned to drive in the late 1940s and 50s so that they would have no further need of public transport. *(A. R. Gault)*

Opposite Top: Grove Park in 1950 with E1 No.1604, sporting a neatly fitting windscreen, but otherwise looking somewhat careworn, about to return to Victoria whilst a rehabilitated E1 waits patiently a short distance off. No.1604 was withdrawn in June, 1951 after a working life of 40 years. *(A. R. Gault)*

Opposite Bottom: A busy scene at Well Hall Road, 16th April, 1952 where an HR2 has just arrived at its destination, short of Beresford Square, Woolwich, whilst No.85, a former East Ham car of 1927, is bound for Central London with an RT close behind. An Austin Devon saloon is trying to ease its way through. *(A. R. Gault)*

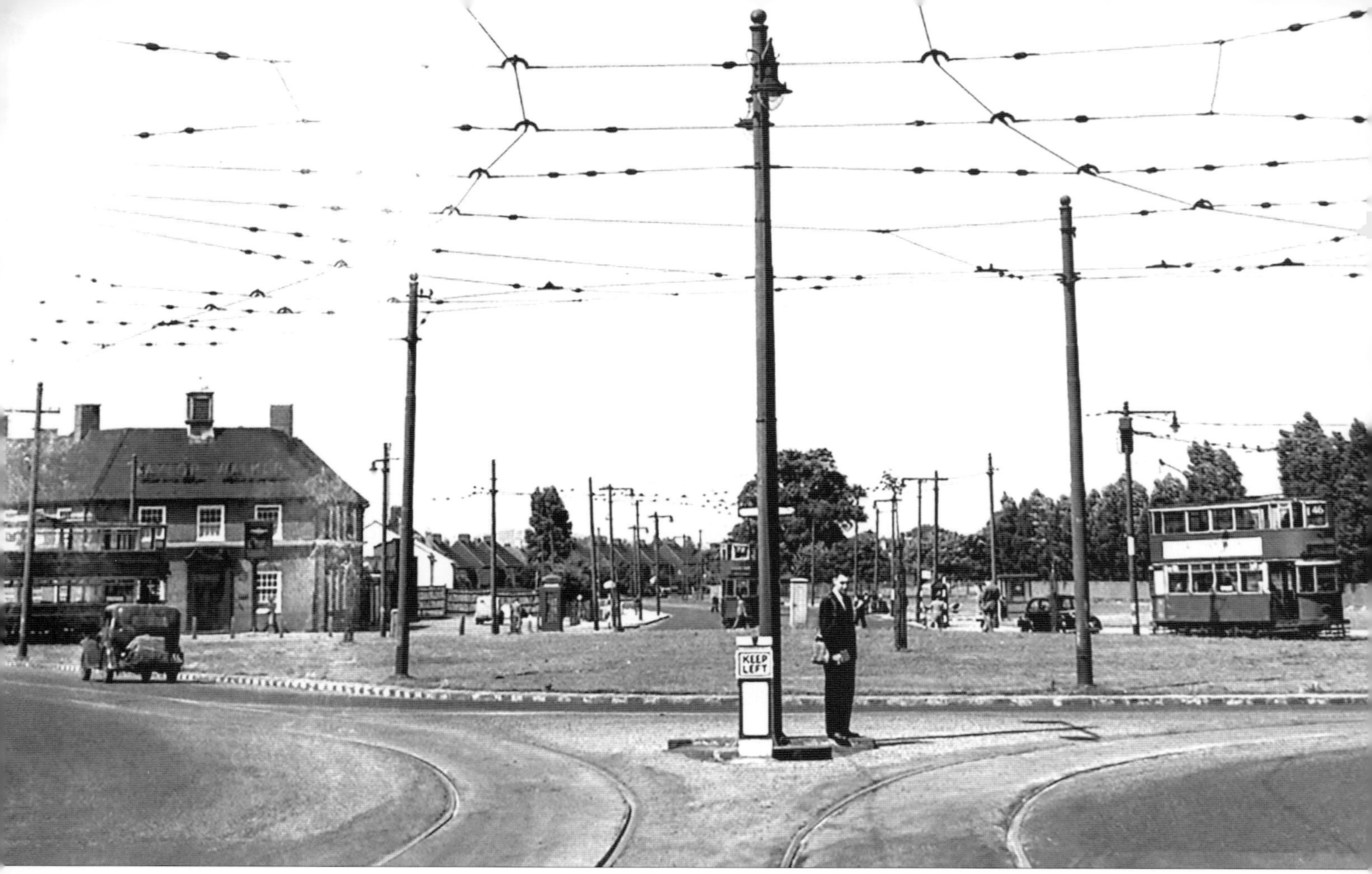
KEEP LEFT

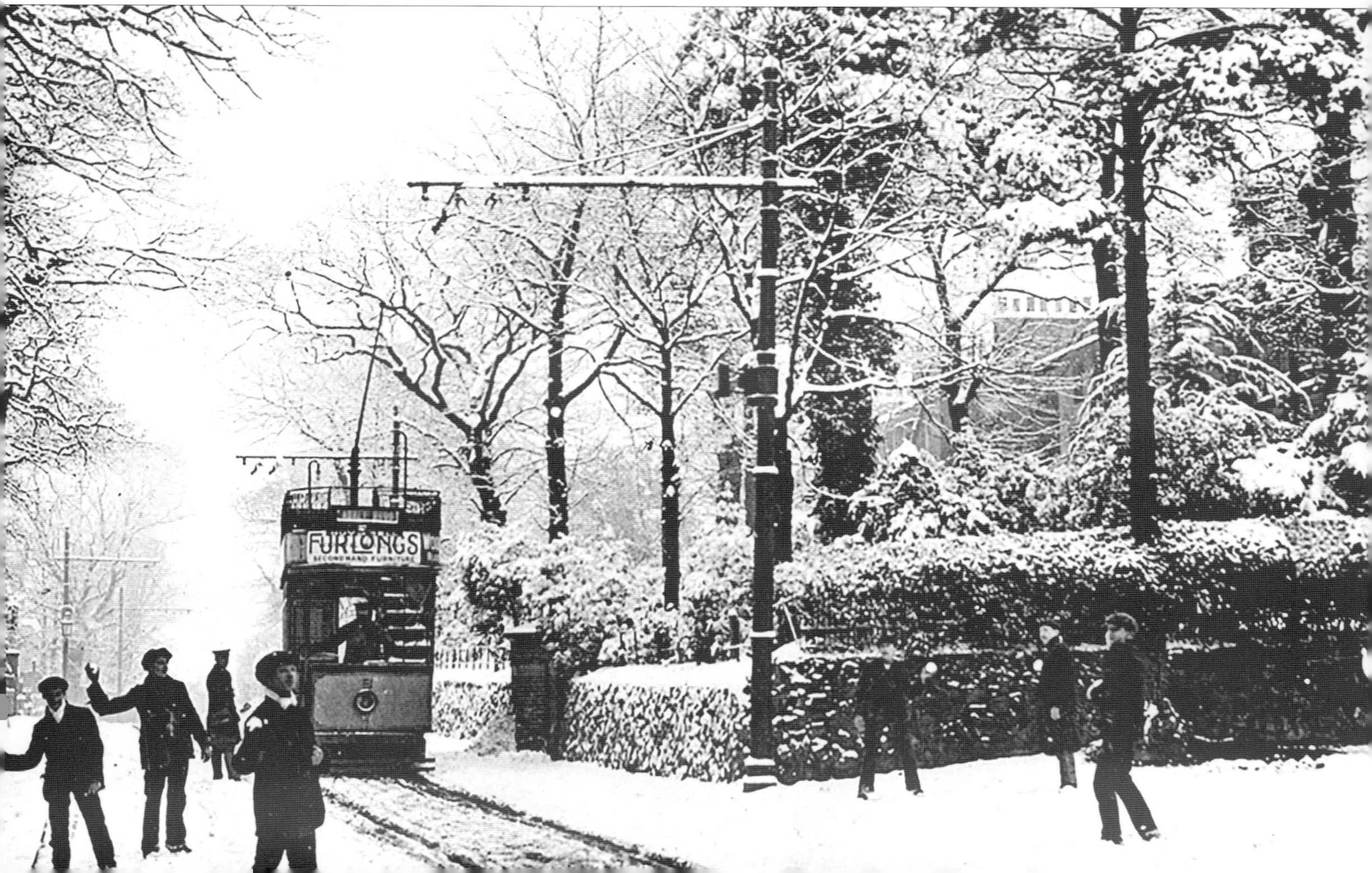
FURLONGS
SECONDHAND FURNITURE

Opposite Top: Circling roundabouts were not what most trams had been trained to do, the tram as a concept predating the roundabout. In the new post- World War Two era of the outer Kentish suburbs we see a couple here doing just that in Middle Park, Eltham, although not for much longer as this photograph was taken on the July day in 1952 when it all came to an end. Hindsight would strongly suggest that this part of the London suburbs was designed for the tram and should never have got rid of it. *(A. R. Gault)*

Opposite Bottom: A very different scene, a very different time in Erith pre-1914 when trams were the latest thing and the populace could expect deep snow every winter. Erith UDC Tramways was the smallest tram concern taken over by London Transport, with 19 four wheel cars and just over four miles of track from Abbey Wood to Northumberland Heath where its trams continued over Bexley tracks to Bexleyheath Market Place. *(Milepost 92½)*

Below: A lady opposite Rehabilitated E1 No.1352 on Southwark Bridge would seem to be trying to make up her mind whether or not to board and head for Telford Avenue depot. Perhaps the non standard route number stencil has her puzzled. *(A. R. Gault)*

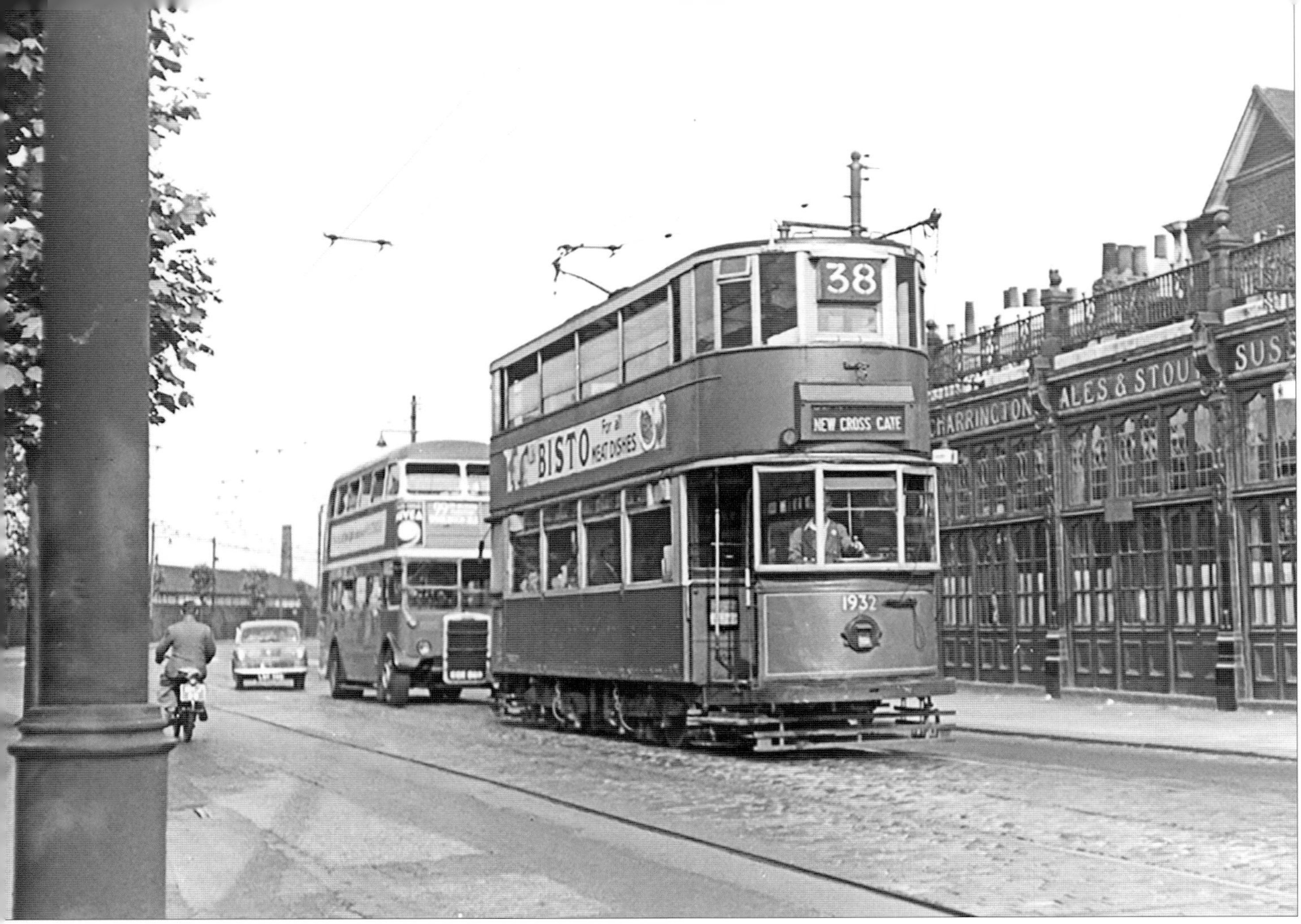
38
NEW CROSS GATE
BISTO
For all MEAT DISHES
1932
CHARRINGTON
ALES & STOU
SUS

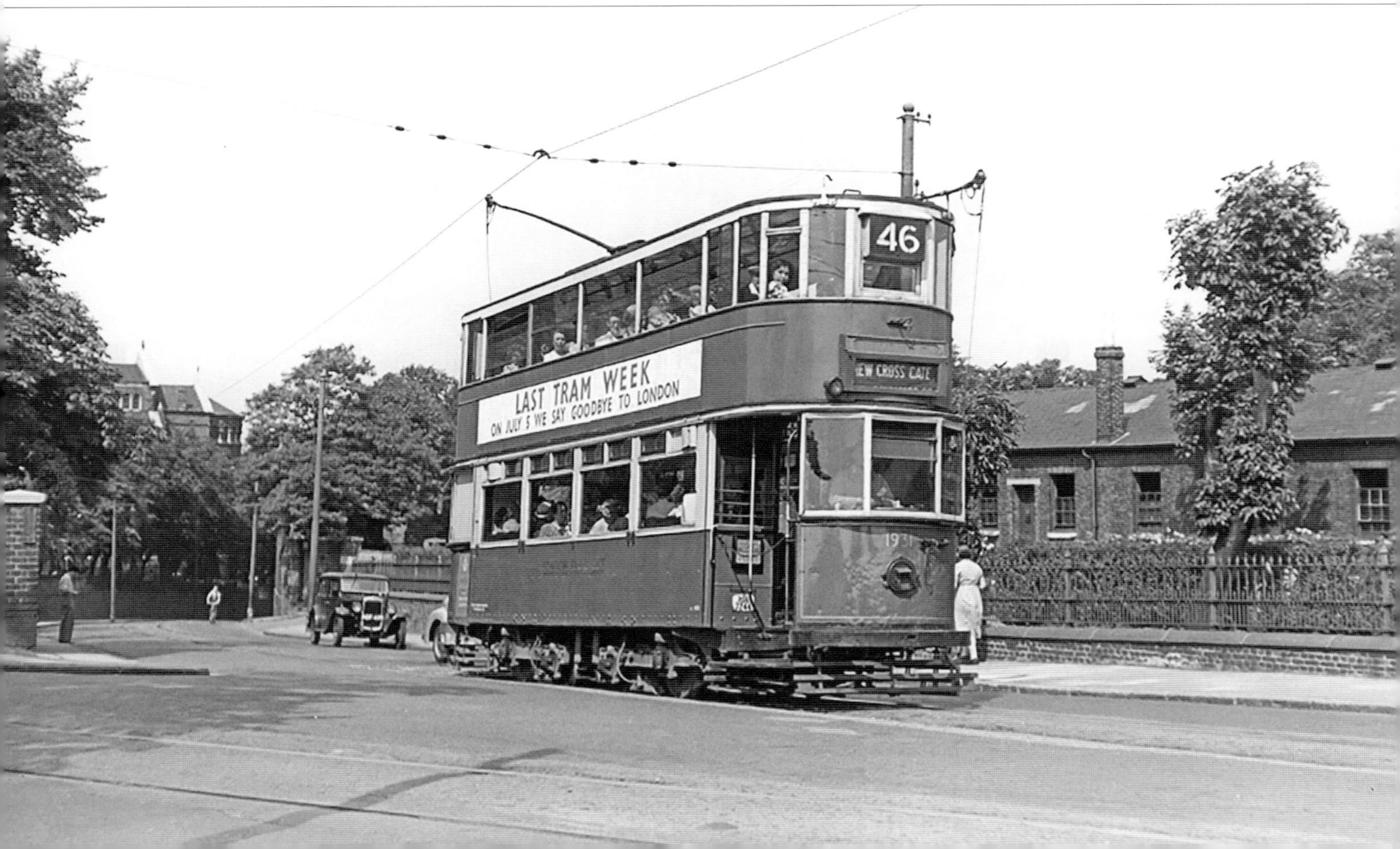
46
NEW CROSS GATE
LAST TRAM WEEK
ON JULY 5 WE SAY GOODBYE TO LONDON
1931

Opposite Top: From Woolwich to Abbey Wood trams and trolleybuses shared the road and the overhead wires. E3 No.1932 is passing the highly ornate Sussex Arms in Plumstead Road, SE18 on 29th June, 1952, ahead of an RTL. The Sussex Arms plied its trade for exactly 100 years, opening in 1862. *(A. R. Gault)*

Opposite Bottom: The sun is out, the passengers are dressed for high summer as E1 No. 1931 climbs Woolwich New Road and joins Grand Depot Road on its 65 minute run to New Cross Gate, 29th June, 1952. *(A. R. Gault)*

Above: Former East Ham car No.92 on its way from Woolwich up Woolwich New Road to Eltham, 29th June, 1952. The bottom deck appears to be full, we can see right through the front entrance to a mother and daughter on the nearside longitudinal seat, but the upper deck seems almost empty. Perhaps as the entire journey lasts no longer than 18 minutes it was hardly worth climbing the stairs. *(A. R. Gault)*

Above: No.2 was a rather special tram, as it should have been with such a distinctive number. It is seen here in Woolwich during, note the poster, the last tram week, July, 1952. No.1 was the most distinctive of all the ex-LCC vehicles, and No.2 made a real effort to follow in its footsteps. Reconditioned in February, 1935, it has sometimes been referred to as the only new tram ever built by London Transport. In fact it was a rebuild and those who did the job had all been employees of the LCC, but it was certainly a step in the right direction which might have led to a whole generation of really modern cars, if the powers that be had not been so ruthlessly anti-tram. The sloping, streamlined upper deck, in particular, was an almost direct copy of No.1's. Lower down its E1/E3 ancestry was obvious. But it was still a great improvement on the penny pinching rehabs which succeeded it. Unlike most one offs it lasted until the very last day of the original tram system. *(A. R. Gault)*

Opposite Top: The complete set, trams, trolleybus, and diesel bus; catch it while you can. Woolwich on 29th June, 1952 with two trams, both E3s, one of them No.1911, a cyclist, D2 trolleybus No.405B, RT 1758 plus the added bonus of a lady in a spotted frock. *(A. R. Gault)*

Opposite Bottom: Tram 217 on route 62 standing at Blackwall Tunnel terminus ready to return to the Embankment (Savoy St). Behind is the southern gatehouse entrance to Blackwall Tunnel. A pair of these gatehouses were erected on either side of the Thames prior to the opening of the tunnel in 1897. They were designed by Thomas Blashill, the Superintending Architect for the London County Council. The northern gatehouse was demolished in 1958, but the southern one still stands and is a grade II listed building. Post war tram route 62 ran from Forest Hill to Victoria Embankment (Savoy Street). After the war no tram routes served this terminus at Blackwall Tunnel. *(Milepost 92½)*

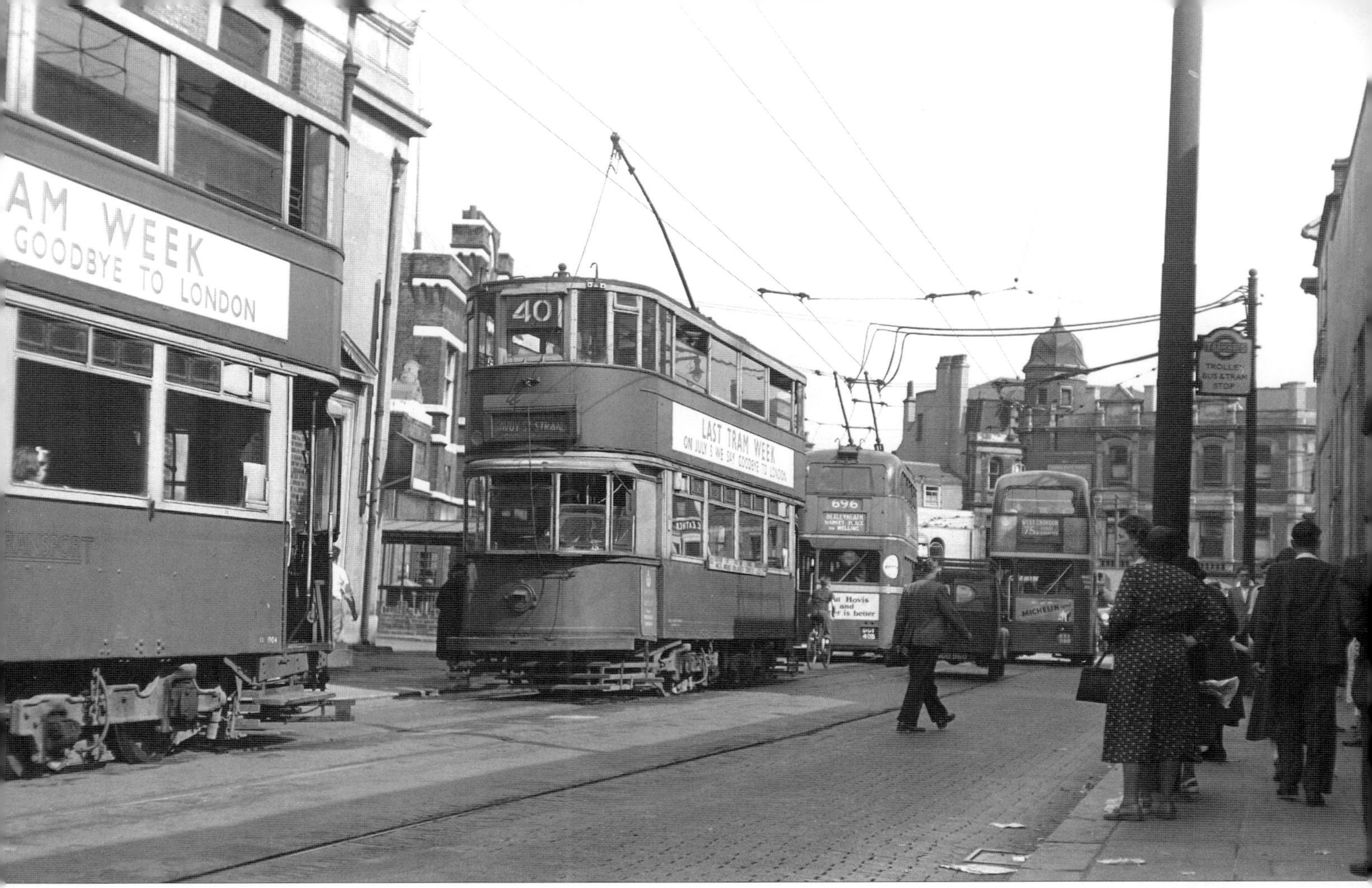
AM WEEK
GOODBYE TO LONDON
40
LAST TRAM WEEK
ON JULY 5 WE SAY GOODBYE TO LONDON
696
TROLLEY BUS & TRAM STOP

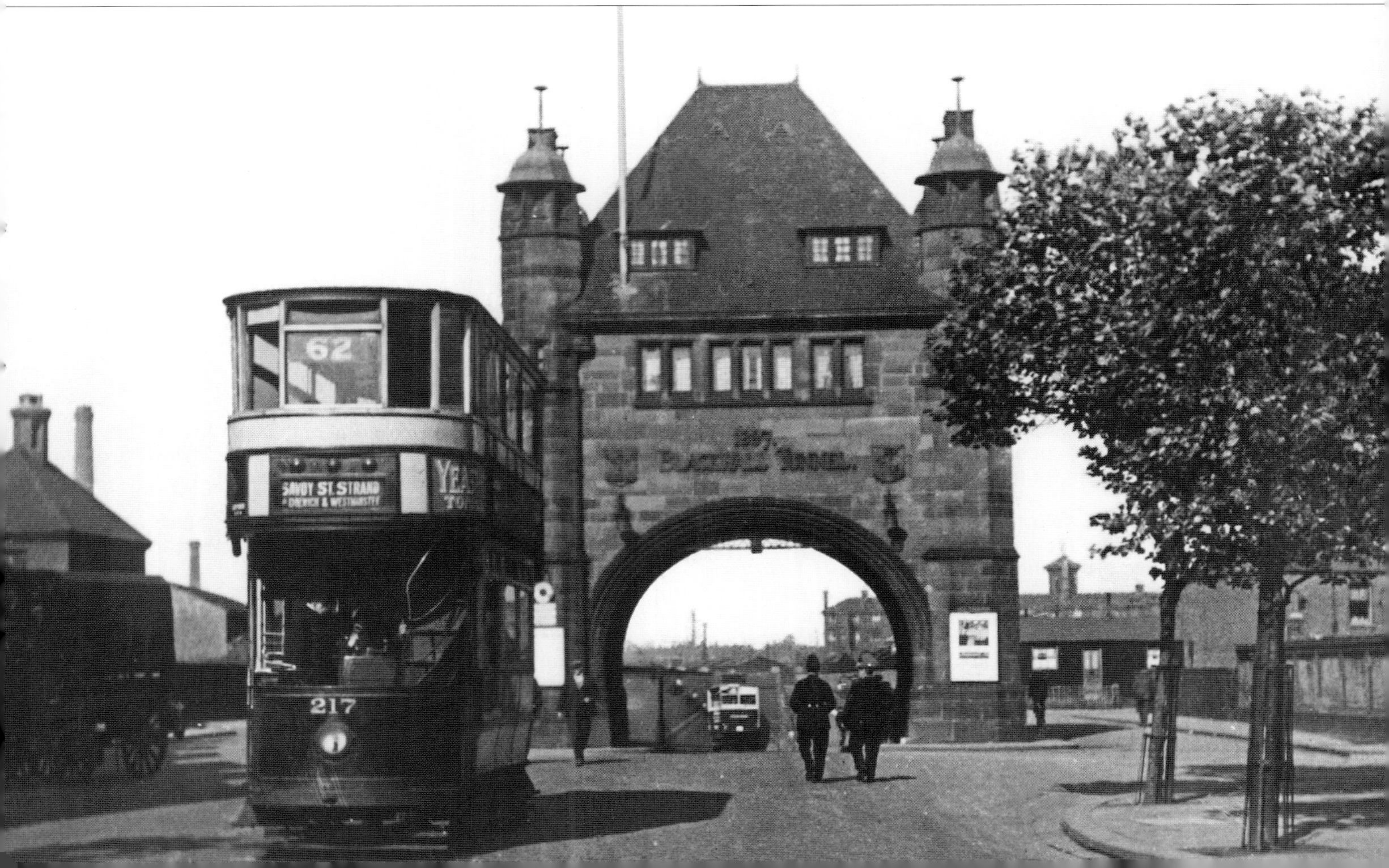
62
SAVOY ST. STRAND
217

Above: E3 No.1936 stands at the Archway terminus of route 35, the most north-westerly point of the compass reached by London trams in post-war years. To the far left and right are gentlemen sporting the long overcoats which were de rigeur wear except for the very hottest high summer days, and this is only Spring in 1952. *(A. R. Gault)*

Opposite Top: E1 No.157 proceeding along the one way Upper Street, Islington on 28th December, 1951 with a passenger ready for a quick get away. The 86 minutes from Forest Hill to Archway was the longest start to finish journey possible on the tram network. *(A. R. Gault)*

Opposite Bottom: Three E3 class trams, the rear one a long way off route 42, are in the centre of a very jolly time involving members of the Light Railway Transport League spreading all over the highway at Highgate on 5th April, 1952. The driver learning to handle the Morris saloon of about the same vintage as the trams must have been very surprised. A trolleybus is keeping well clear in the distance. *(A. R. Gault)*

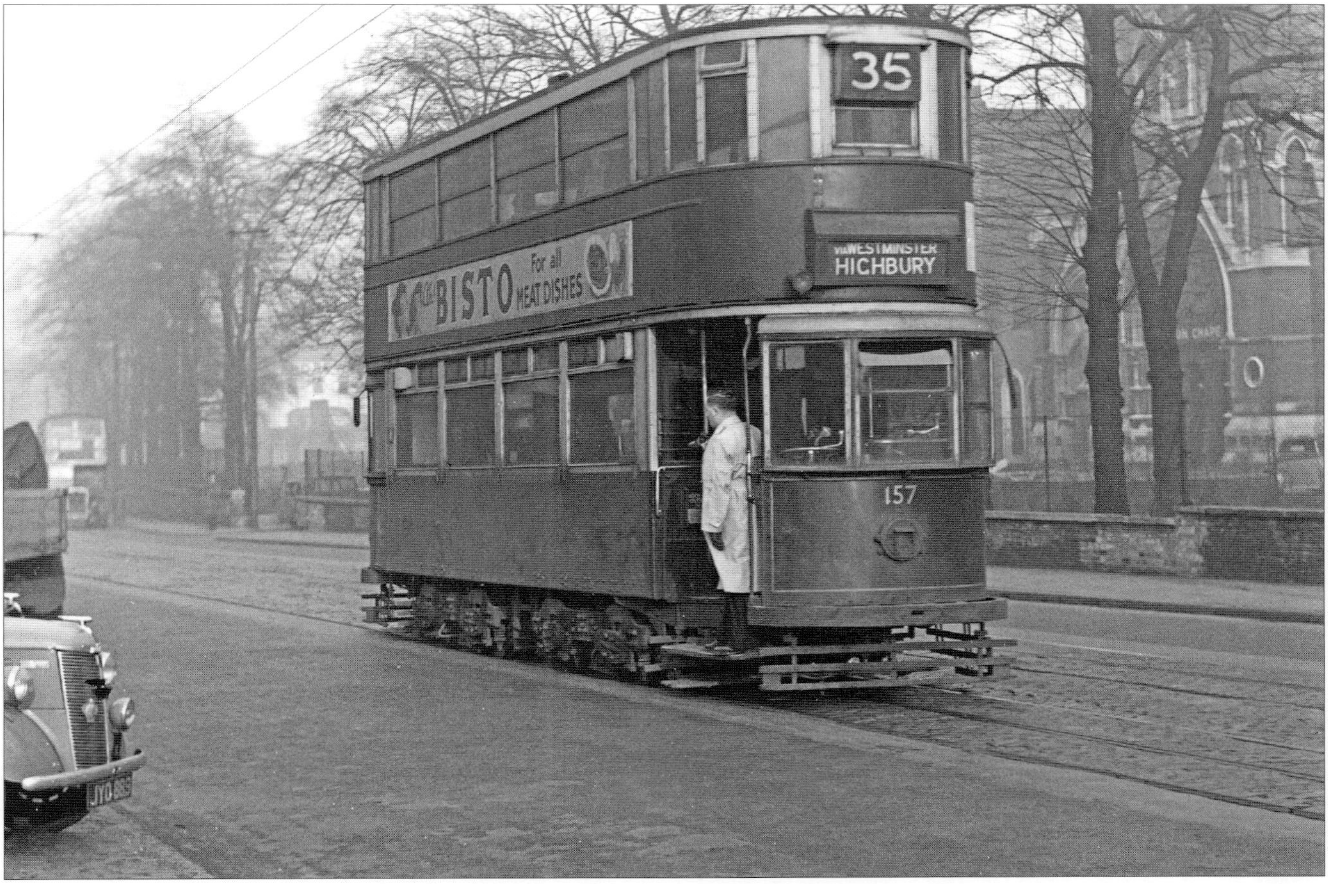
35
viaWESTMINSTER
HIGHBURY
BISTO
For all
MEAT DISHES
157

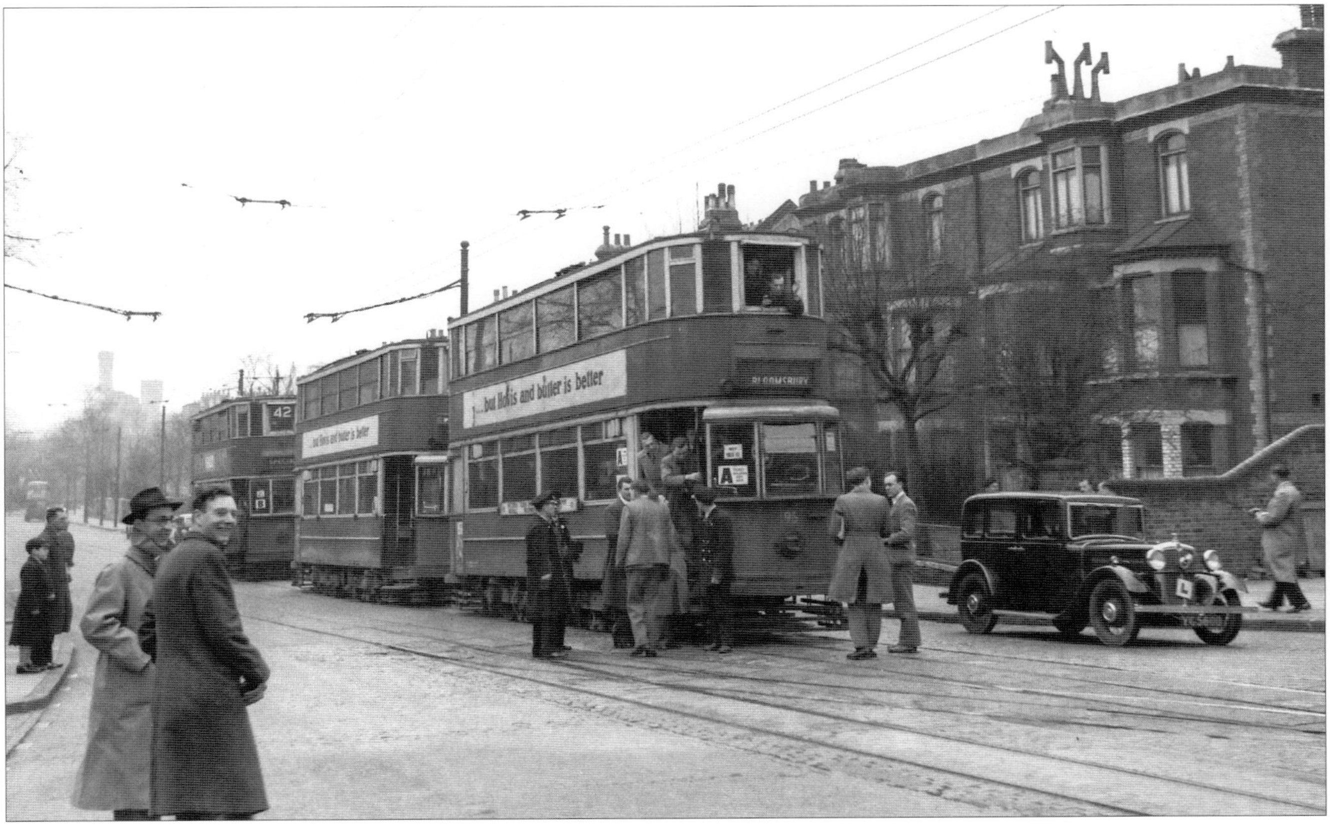
BLOOMSBURY
...but Hovis and butter is better
42

Above: E1 No.1782 has just turned off the Embankment and gained Blackfriars Bridge on its way to Wimbledon, 6th January, 1951. A girl has the favoured position upstairs by the very front where you could sit with your nose pressed against the window, if you so desired. Beyond is a British Railways mechanical horse and in the far distance an RT. *(A. R. Gault)*

Opposite Top: Closely related, in more ways than one, was route 70 to route 68. It too was worked by 1930 vintage E1s, such as No.570 seen here on 10th July, 1951, and was the only route to use the Tooley Street terminus alongside London Bridge station, just as route 68 operated beside the not very distant Waterloo. *(A. R. Gault)*

Opposite Bottom: One of the magnificent Felthams looking very sad and down at heel at the Southwark terminus of route 6 on what is clearly a hot, sunny day c1949. *(A. R. Gault)*

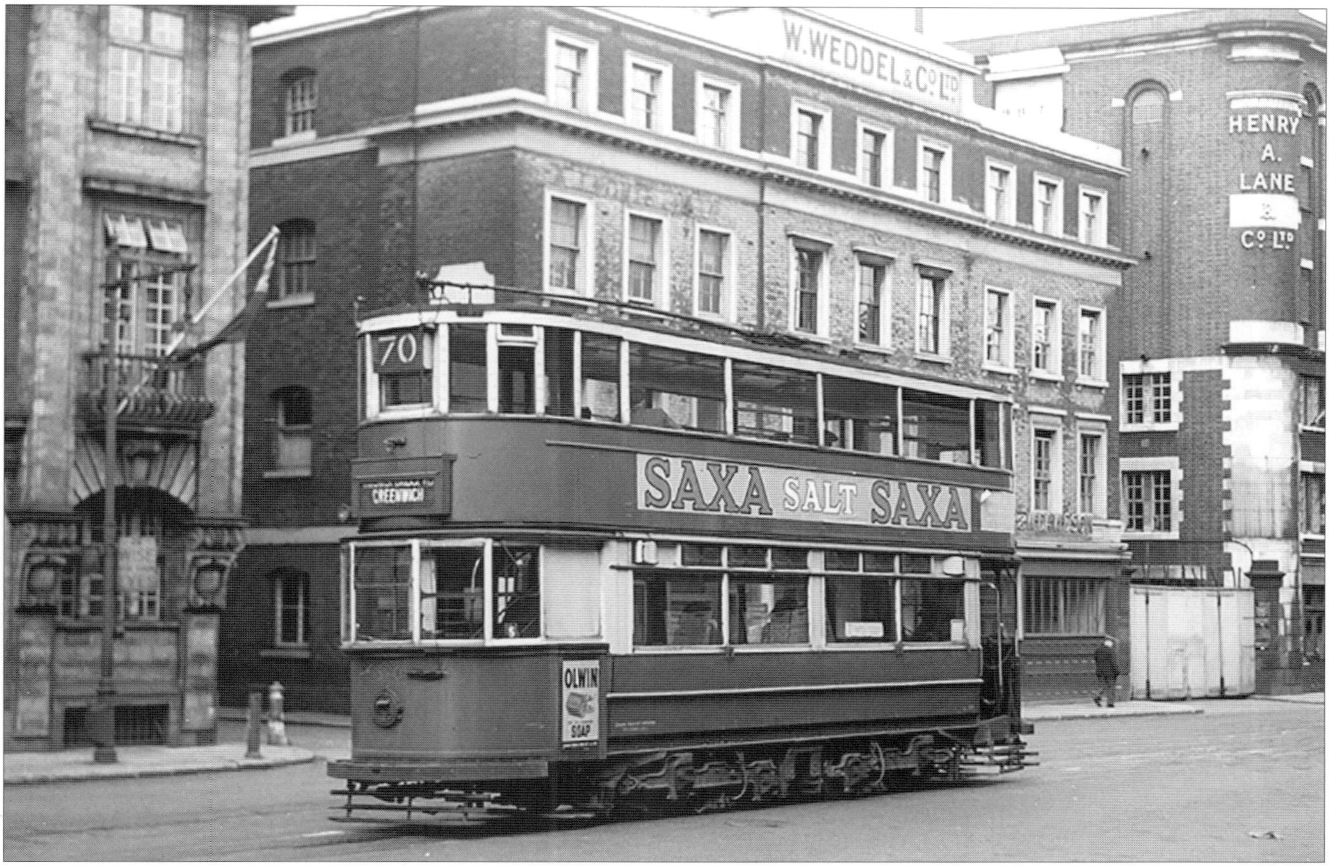
W.WEDDEL & Co Ltd
HENRY
A.
LANE
&
Co Ltd
70
GREENWICH
SAXA SALT SAXA
OLWIN
SOAP

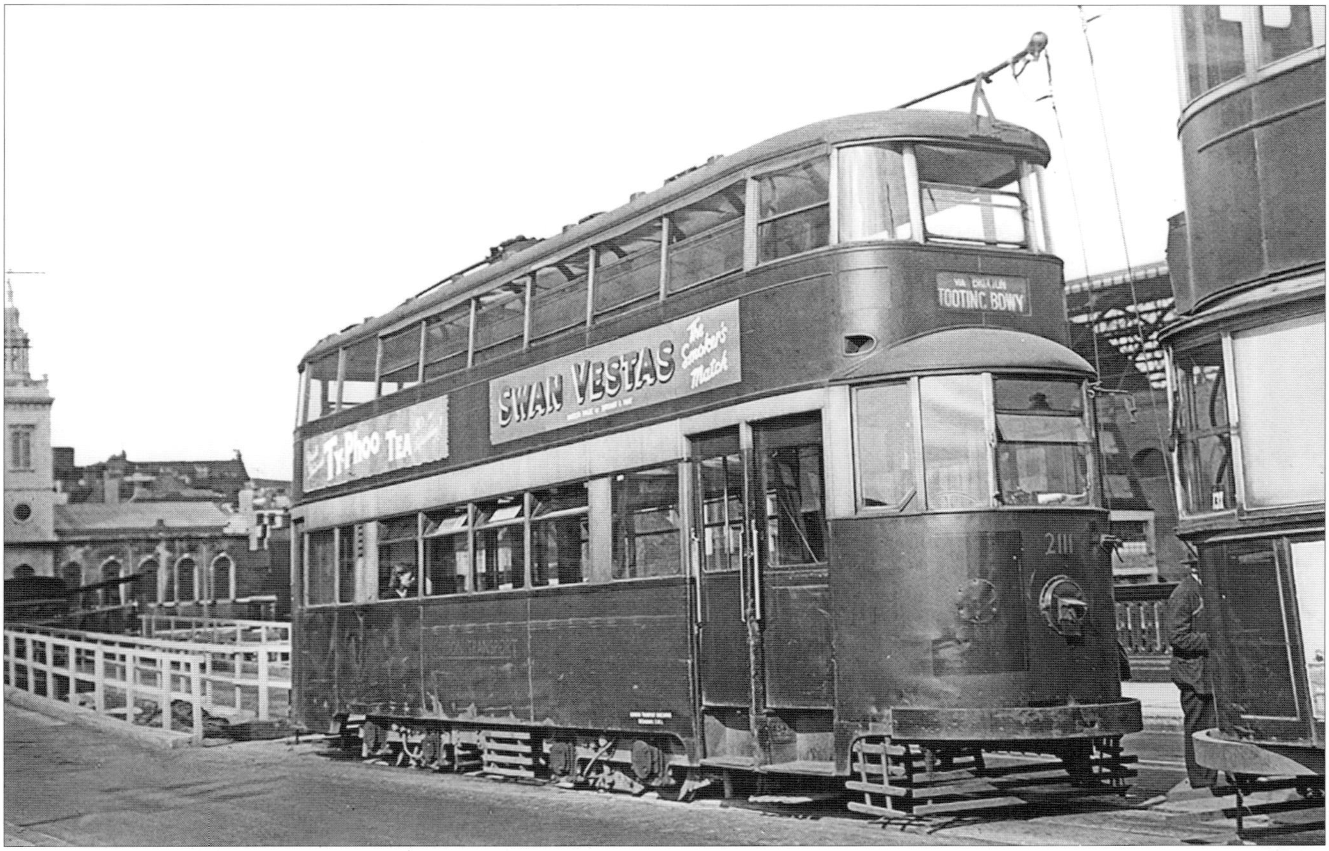
TOOTING BDWY
SWAN VESTAS
The Smoker's Match
TyPHOO TEA
2111

Top: A rather more cared for rehabilitated E1, No.1388, again on a warm, perhaps even hot, sunny afternoon in 1949. The City end of Southwark Bridge had been bombed in the war and the temporary structure seen by where the three ladies are crossing sufficed until the last tram had quit the bridge for ever. The via board still in place beside the entrance on the E1 was an unusual feature on the rehabilitated cars in post-war days. *(A. R. Gault)*

Bottom: Definitely neither a warm nor a sunny day on Southwark Bridge on this occasion. E3 No.1938 will shortly be setting off for Norwood. Hundreds, if not thousands, of working horses could still be found on the streets of London and its suburbs as long as trams were still about, and for a while afterwards. *(A. R. Gault)*

Opposite: An unusual view of the centre of Kingston where four main roads converge. London united tram 277 appears to be leaving Clarence Street on the left foreground. On the right foreground the tracks are coming from Eden Street. Ahead the tracks diverge, to the left along Richmond Road, past the train station, and to the right along London Road. *(Milepost 92½)*

KINGSTON ON-THAMES. Clarence Street. — LL

Above: Two variations on the E1 theme at Southwark on 6th January, 1951. Leading is No.296, a Brush/Hurst Nelson car, delivered to West Ham Corporation in 1929, whilst behind is former LCC No.1787, also built by Brush/Hurst Nelson in 1922. Clearly the office buildings behind have survived the Blitz, although no doubt they would have suffered some damage. *(A. R. Gault)*

Opposite Top: Feltham trams 2098 on route 22 to Tooting is turning out of the Embankment on to Westminster Bridge in the early part of 1949. *(Milepost 92½)*

Opposite Bottom: Finally on this rather bleak January day in 1951 on Southwark bridge No.206, an E3 of pure LCC design but actually put into service by Leyton Corporation, is about to set off for the delights of Tooting Broadway. Southwark bridge was always one of the least busy crossing points of the Thames in central London, with or without trams, and that is still the situation. 6th January, 1951. *(A. R. Gault)*

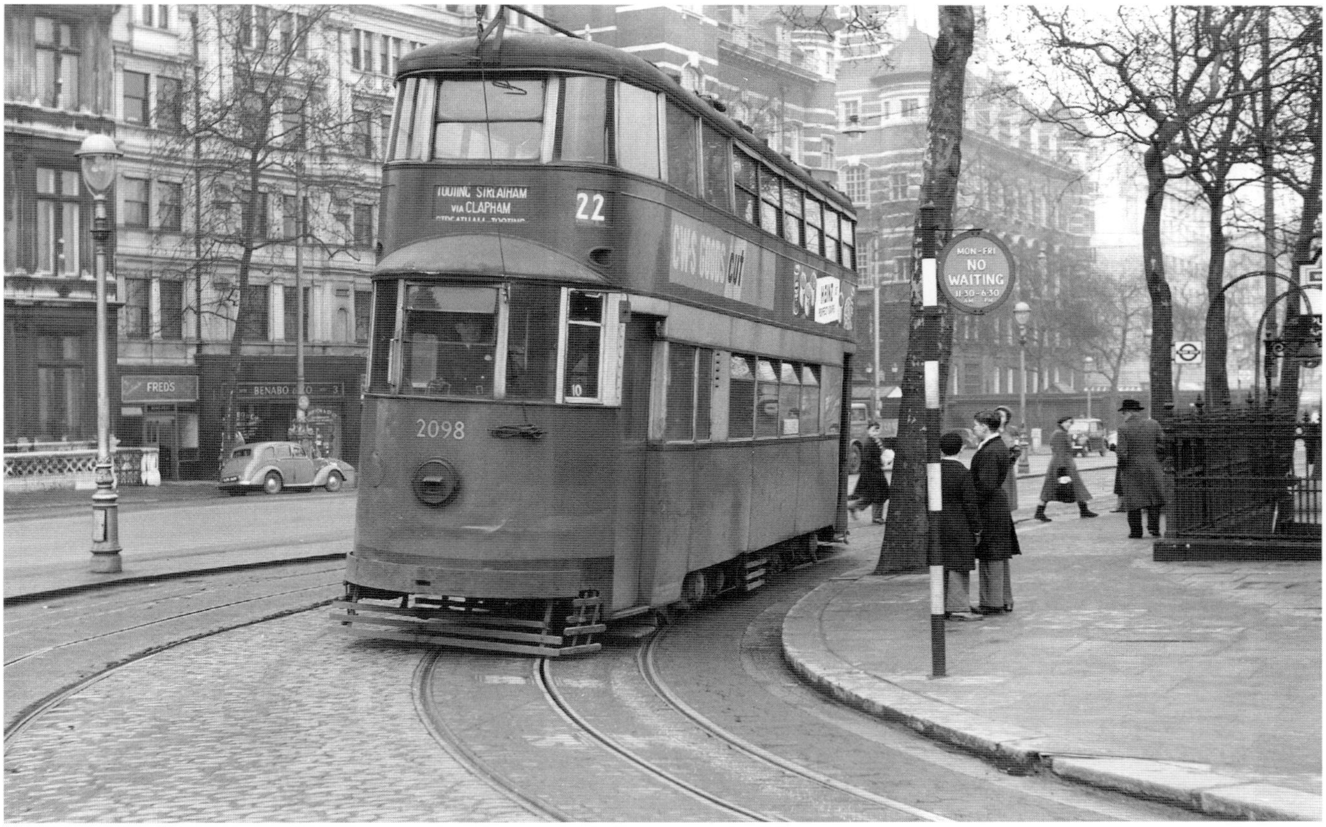
via CLAPHAM
22
2098
MON-FRI
NO
WAITING
FREDS

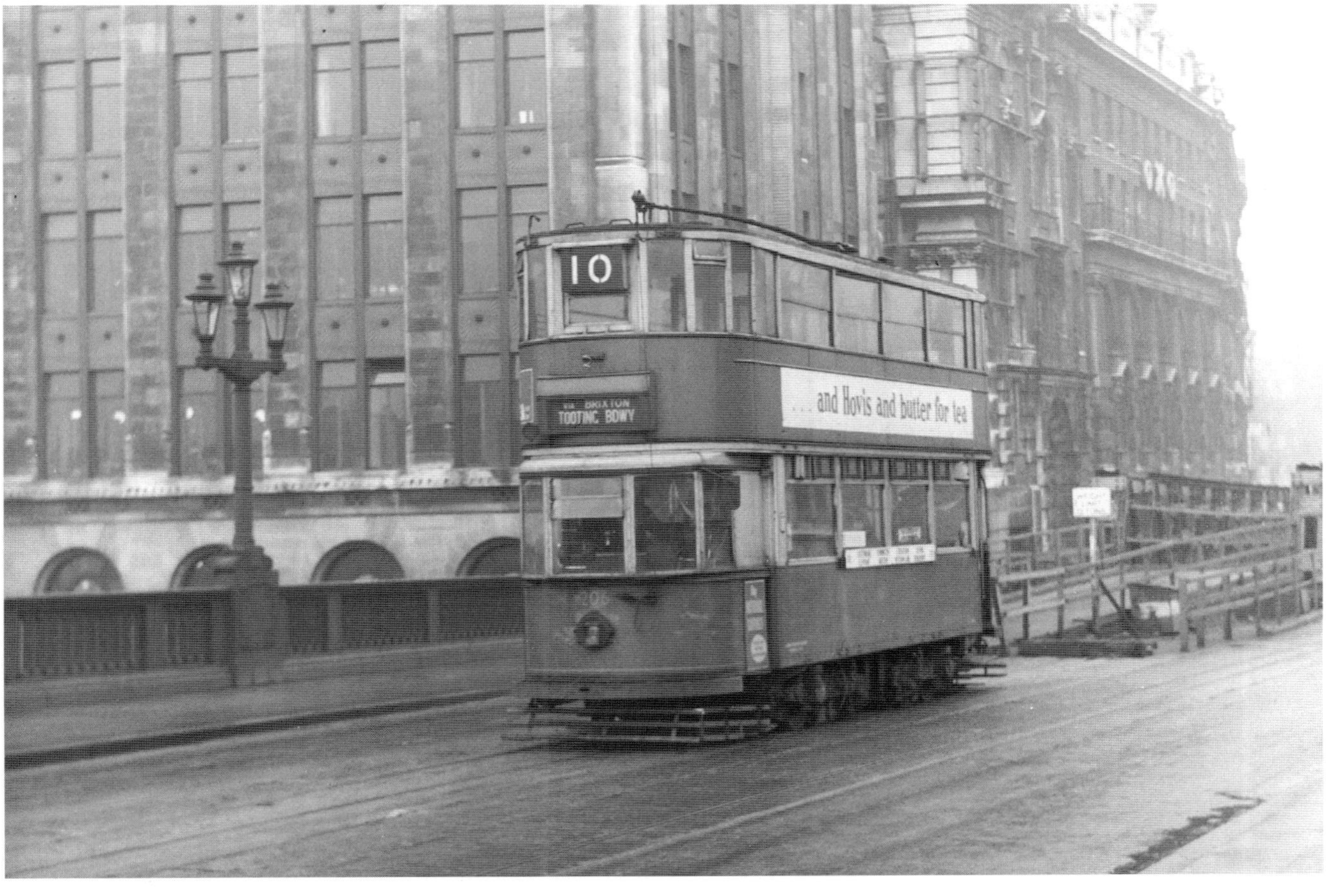
10
BRIXTON
TOOTING BDWY
... and Hovis and butter for tea

Above: Moving up river we come to Blackfriars Bridge. It's still 6th January, 1951, the sun hasn't appeared but E1 No.1829, with a dented front (not unusual in post-war years, possibly as the result of a bump in the night in an unlit wartime street), has learned long ago in its 29 year career to expect nothing better in January, and there is almost certain to be worse, heavy snow for instance being more than possible before March is out. As it happens the winter of 1950/1 was mild compared with that at the beginning of 1947 when so fierce was the cold that the country almost ground to a halt, coal was in short supply as was electricity, and the Minster of Fuel, Emanuel Shinwell, received death threats! Equally awful were November and December, 1952, by which time trams were no more in London, with choking smog enveloping London, causing the death of 1,000s of people and bringing traffic to a standstill. Although smoke from homes, factories and steam locomotives was the chief culprit, diesel powered buses contributed a measurable amount: trams never did. *(A. R. Gault)*

Opposite top: No.1144, one of several E1s rehabilitated in the 1930s, heads south over Blackfriars on its way to Grove Park, 30th December, 1951. The structural changes made on rebuilding were minimal, and no E1Rs survived until the last day in July, 1952. However one, No.1622, rose, metaphorically, from the ashes when it began once again carrying passengers at the National Tram Museum at Crich in 1997, although strictly speaking only the lower deck had survived, first as a caravan on Hayling Island, later in an orchard in Hampshire. *(A. R. Gault)*

Opposite Bottom: No.391, one of the former Croydon cars which briefly escaped the fate of most of its fellows when the Croydon area trams were withdrawn. It was the only one with flush sides. No.391, seen on Blackfriars Bridge, is heading south, 30th December, 1951, serving a few months longer from New Cross depot. *(A. R. Gault)*

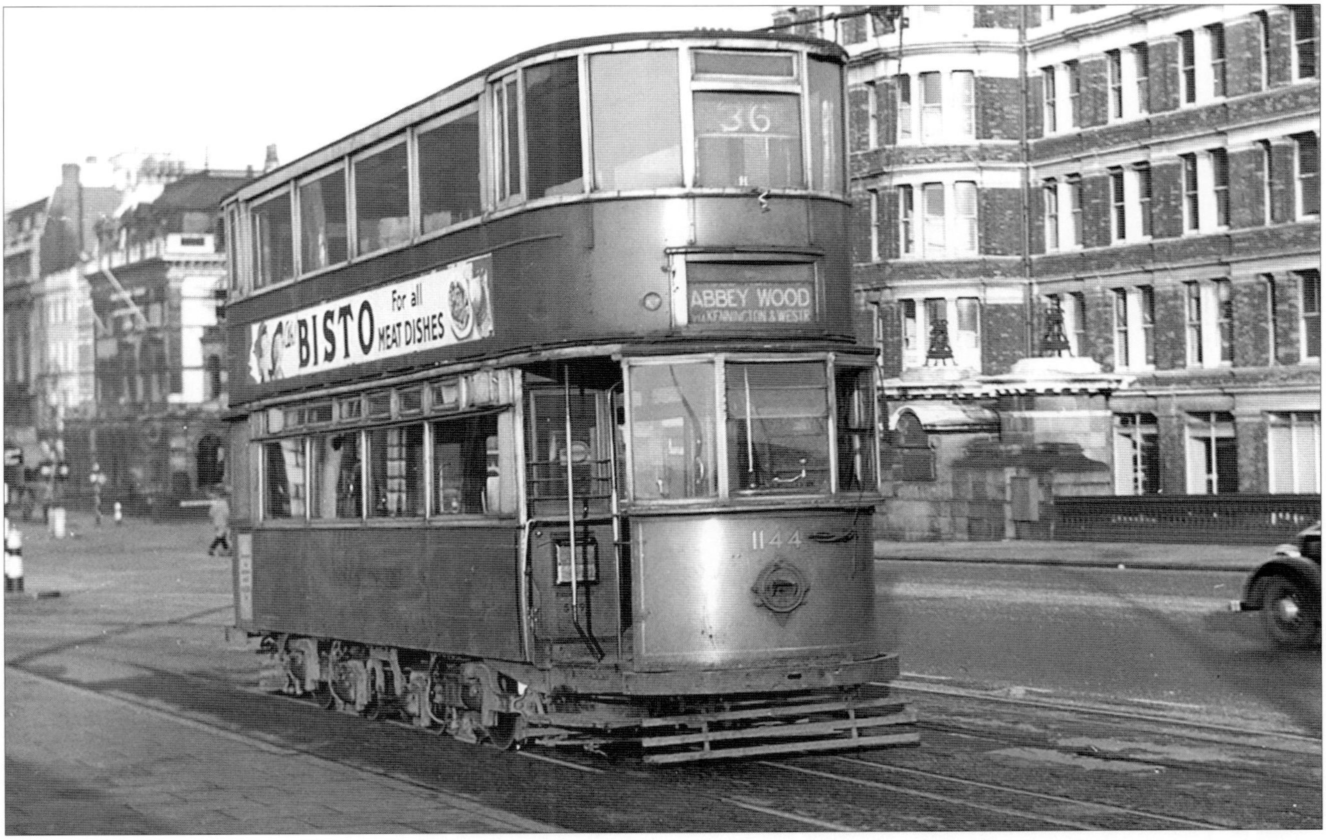
36
ABBEY WOOD
BISTO
For all
MEAT DISHES
1144

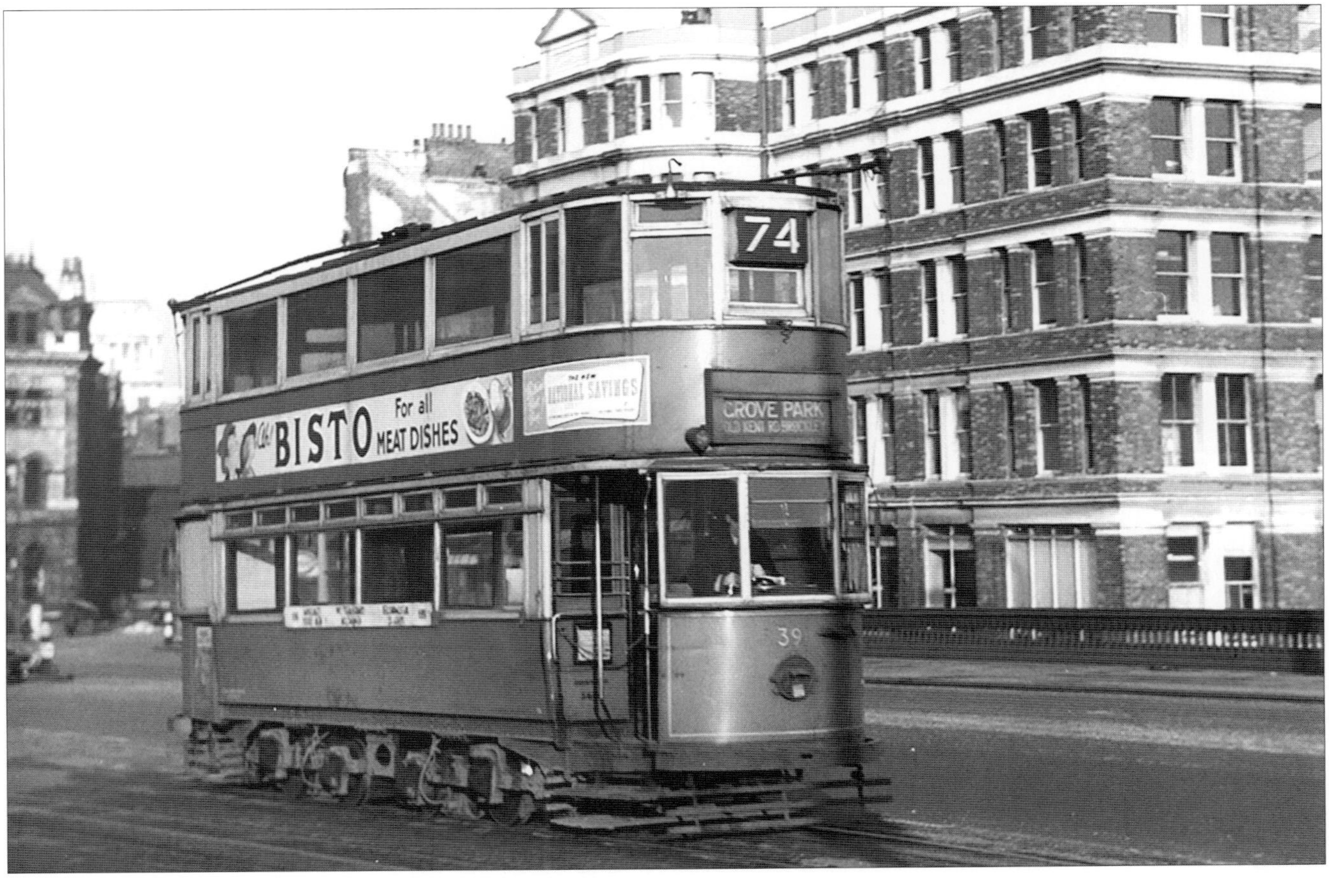
74
GROVE PARK
BISTO
For all
MEAT DISHES
NATIONAL SAVINGS
39

Although Westminster Council did not like the notion of nasty, noisy tram cars full of the working class sullying their streets they had, reluctantly, to allow them on Westminster Bridge and thus the tram reached the very heart of this tourist hot spot. A former West Ham E1 car is passing tourists gazing over the river, its passengers ready to dismount and join them, as it approaches the Embankment, under the shadow of Big Ben, on a summer's day in 1951.
(A. R. Gault)

A few yards further on and we have reached the end of Westminster Bridge. A Peckham Rye bound HR2 has just turned off the Embankment, it is accompanied by a motor cycle combination and a Ford 8 van whilst a tram replacement RT is heading along the Embankment. September, 1950 marked the first time motor buses had ventured on to what had been until then the trams' monopoly. *(A. R. Gault)*

Tram E3 No 165 on route 66 (Victoria-Forest Hill) passing the offices of the Temperance Society in Camberwell on its way to Victoria. The building is still there and is listed. Ahead is E1 1798 on route 54 (Victoria – Grove Park) also heading towards Victoria. *(A. R. Gault)*

The passing policeman appears to be doffing his helmet to E3 No.1939 as he proceeds, in accordance with instructions, along Dove Road, Canonbury, 4th April, 1952, past a row of early Victorian terraces. Very likely the patch of grass in the right foreground is a not yet built upon bombsite, a familiar sight in London well into the 1950s. Today Dove Road is utterly changed, flats occupying much of it, and parked cars everywhere. Along with tram route 33 it was also served by trolleybus routes 581 and 677. *(A. R. Gault)*

Top: Passers-by jostle where Westminster Bridge meets the Embankment with the Houses of Parliament beyond the brand new RT working tram replacement route 109 and a not yet banished rehabilitated E1/R tram in the summer of 1951. *(A. R. Gault)*

Bottom: An HR2 stands at the Victoria terminus of route 58 in 1950 before setting off for Greenwich. The side view of this sturdy, if less than pristine vehicle, dating from 1930 would not have looked so very different to an Edwardian standing in the same spot 40 or years earlier. Modern touches are the flush, metal sides and windscreens protect the driver, but there are the same number of windows, boards still indicate the suburbs it will be serving, the number is a stencil, the platform upon which the driver and conductor carry out their duties are open to the elements, there is no provision for the former to sit down. *(A. R. Gault)*

Even longer was the 86 minute journey of the 35 from Forest Hill along the Embankment, through the Kingsway Subway and way up into the north London suburbs terminating at Archway Northern Line tube station, Highgate. A newly overhauled E3 No.1950 is well on its way and about to enter the subway sometime in 1951. *(A. R. Gault)*

In this 1951 photograph some 100 yards to the west of the previous picture another E3, No.1945, is heading along the Embankment and about to pass under Hungerford railway bridge with a tram replacement RT on the immediate right, another E3 beyond and a c1937 Morris 8 saloon also in view. The metal bodied E3s were the only trams, officially, allowed in the Kingsway Subway. *(A. R. Gault)*

Feltham No.2068 shows how perhaps unexpectedly easy it was for these much longer than standard trams to negotiate sharp curves as it leaves the Embankment and heads on to Westminster Bridge. It is on a short working to Brixton Hill depot, a building which still exists, with all its tracks complete and is now home to Arriva buses. Beyond, facing the Houses of Parliament, is St Thomas' Hospital, founded in the City of London in the twelfth century, and much rebuilt over the years. *(A. R. Gault)*

Above: Retracing our steps past Westminster and Hungerford bridges we find ourselves standing underneath Waterloo Bridge, another central London one which never carried trams but under which, as can be clearly seen, they burrowed. This was London's newest bridge, plain and elegant, constructed of concrete beams with Portland stone facing, designed by none other than Sir Giles Gilbert Scott, he of Liverpool C. o. E. cathedral fame, the second largest in Europe after St Peter's Rome, and the GPO telephones call box, now transformed from somewhere from which you could make a phone call to a subject for the mobile phone cameras of tourists. Waterloo Bridge was virtually completed, and in use from 1942, but finishing touches were not added until 1945. *(A. R. Gault)*

Left: Holborn station, 4th May, 1952. Station it surely was, this underground stopping place vastly grander than a mere tram stop. It was to all intents and purposes an underground railway station and one spacious enough to accommodate a double deck tram. A trolleybus with two special entrances/exits was designed in the hope that it could replace a tram in the subway but when it was towed down and tried out, to no-one's surprise it proved impracticable. Looking back – how wonderful is hindsight – it would have been vastly more sensible to retain the tram routes which connected south and north London through the subway, until the time came when, as in so much of Europe, North America and beyond, the tram was once again seen as the perfect solution to urban street travel. Most authorities in south-east London remained, right to the end, in favour of retaining trams, but London Transport, obsessed with the virtues of the diesel bus, knew better. *(A. R. Gault)*

Under Waterloo Bridge, 5th April, 1952. E3 No.1941 is swinging around off the Embankment and into the subway whilst another E3 on route 35 is emerging. Behind the tree is a canvas hut which provided protection for the inspector positioned there. In earlier days this meagre shelter would have been regarded as an unnecessary luxury, just as were windscreens for tram drivers. *(A. R. Gault)*

North and south bound E3s pass on the incline at the northern entrance to the subway, whilst above an STL heads northwards, 5th April, 1952. Apart from the absence of the trams little has changed today, the tracks still being in place, metal gates now keep out trespassers and any last tram which might have found itself marooned and abandoned at Manor House or Archway. *(A. R. Gault)*

Conversation piece at the top of the incline with the driver of E3 No.1981 seemingly in no hurry to move off, 5th April, 1952, the last day of trams through Kingsway Subway. A group of enthusiasts, at least one with a camera, stand on the left. Opposite, awaiting the traffic lights to turn green and allow them to set off up Southampton Row, are a 1939 vintage taxi and a sturdy looking Albion truck. *(A. R. Gault)*

The south end of the Subway, 5th April, 1952 with two maintenance men at work on the conduit. The lookout, note his flag planted between the tracks, is keeping an eye on the emerging 33. Just visible is an RT on a tram replacement route. *(A. R. Gault)*

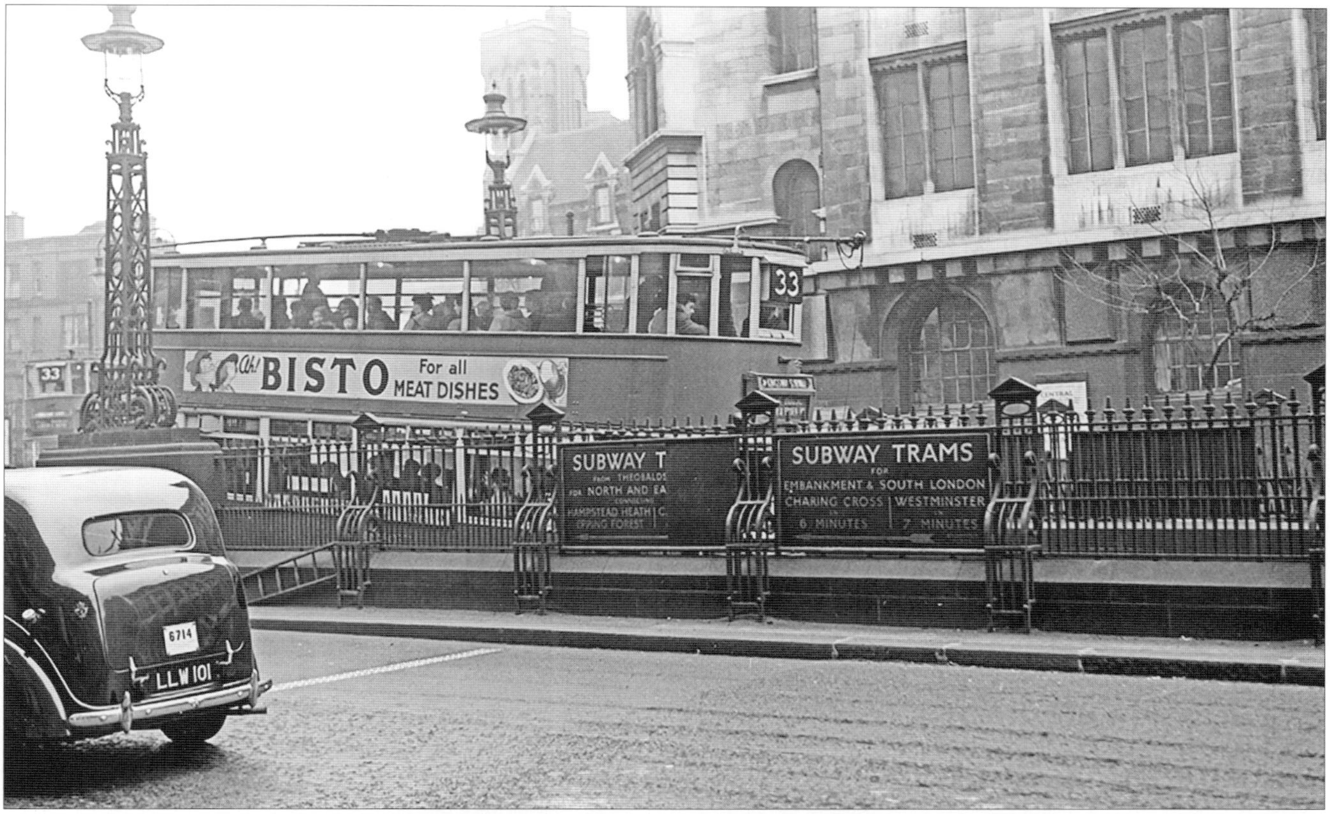

Top: A well filled 33 climbs out of Kingsway Subway, 5th April, 1952, whilst in the distance can be glimpsed a southbound tram in Theobalds Road. It was a long time since some of the destinations featured on the advert attached to the railings, for instance, Hampstead Heath and Epping Forest, could have been reached by tram. *(A. R. Gault)*

Bottom: An evocative night scene at the southern entrance to Kingsway Subway. Two trams on the Embankment have stopped long enough for their image to be captured, somewhat faintly, in this time exposure photograph. Bus routes which replaced the subway trams used Kingsway and Southampton Row. The Embankment end of the tunnel was eventually converted into motor traffic use, much of the rest remains as it was after the last tram had passed through. *(A. R. Gault)*

Above: Victoria was the closest trams got to the West End. Immediately behind E1 No.1005 is the Victoria Palace theatre, at that time, in 1951, the home of the Crazy Gang, an enormously popular comedy troupe, Flanagan and Allen being its best known members. It was Bud Flanagan who sang the signature tune of TV's Dad's Army, which nearly everybody claimed to remember from the World War II years. It was actually written well after, especially for the TV show. *(A. R. Gault)*

Opposite Top: Another E1, No.1546, has moved up behind No.1005, trapping it so that it is a situation of first come, last served. Both are working route 54 for Grove Park and will presumably closely follow each other for the next hour and two minutes. Alongside, heading into Victoria Street, is a station wagon, fitted with a very boxy looking body, a type of vehicle becoming very popular in the post-war years. *(A. R. Gault)*

Opposite Bottom: The classic location if you wanted a photograph of a London tram at work. Taken on 2nd July, 1952, three days before the end, E3 No.184 is heading over Westminster Bridge for south-east London, which always seemed the most natural habitat of the London tram and where the last six routes, from New Cross and Abbey Wood depots, worked. By this date virtually all the original LCC E1s, dating back to pre-World War One days, had met a fiery end, E3s, HR2s and more modern ex- municipal versions of the E1 carrying out the last rites. The temporary keep left sign will not be needed once the last tram has passed. *(A. R. Gault)*

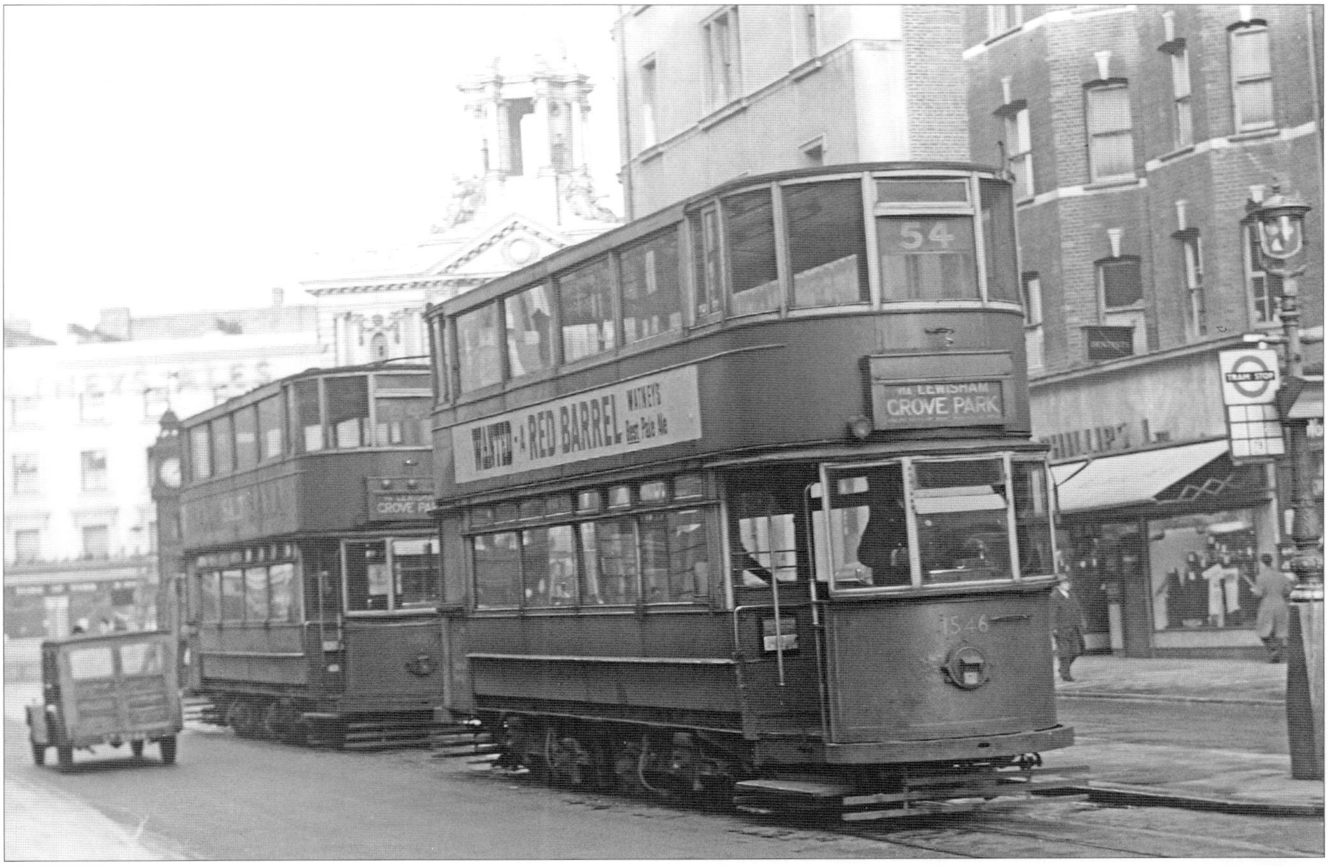
54
WANTED - A RED BARREL

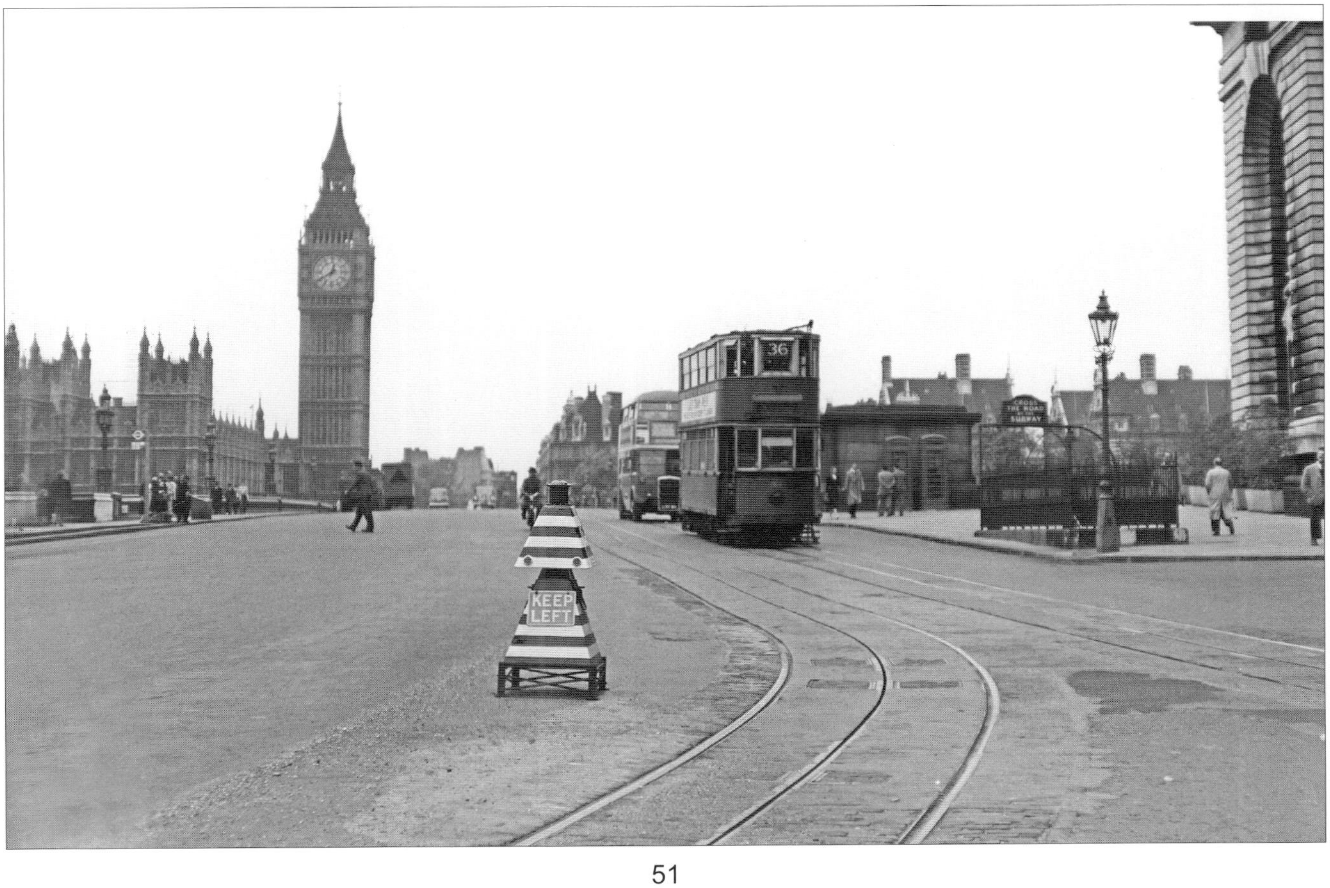
36
KEEP LEFT

Part II
Trolleybuses

Although the travelling public was at first delighted with the trams provided for it by the LUT, the company had over extended itself, the trams began to wear out and the track deteriorated and the money was not there to remedy the situation. One solution was the trackless trolley, which could draw right up the kerb for passengers to board and alight, ran on pneumatic tyres and was thus quieter and more comfortable. What was soon renamed as the trolleybus began operation between Twickenham and Teddington in May, 1931. The scene at Twickenham in May, 1931. A fleet of 60 vehicles with chassis by AEC and bodies by the Union Construction Company of Feltham entered service and was soon working some 17 miles in an area based on Kingston. *(Author's Collection)*

Opposite: For reasons which no-one seems able to explain the original vehicles acquired the nickname Diddlers. Growing increasingly old fashioned they nevertheless survived the Second World War. All were withdrawn in 1948. No.1 was preserved by London Transport and is seen here outside the impressive Fulwell depot, from which it had once operated. *(A. E. Bennett)*

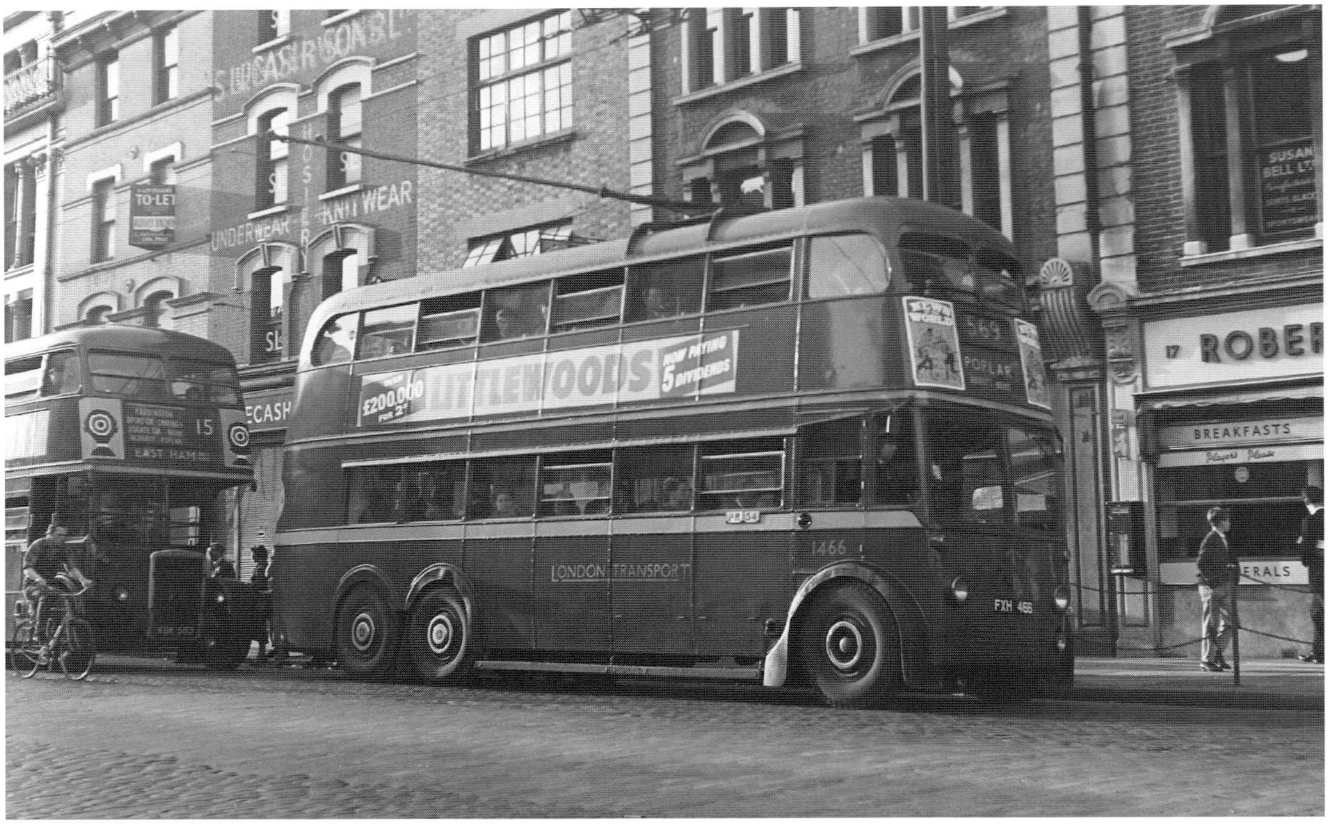
LITTLEWOODS
NOW PAYING
5 DIVIDENDS
£200,000
569
POPLAR
LONDON TRANSPORT
1466
FXH 466
15
EAST HAM
KNITWEAR
TO-LET
17 ROBER
BREAKFASTS

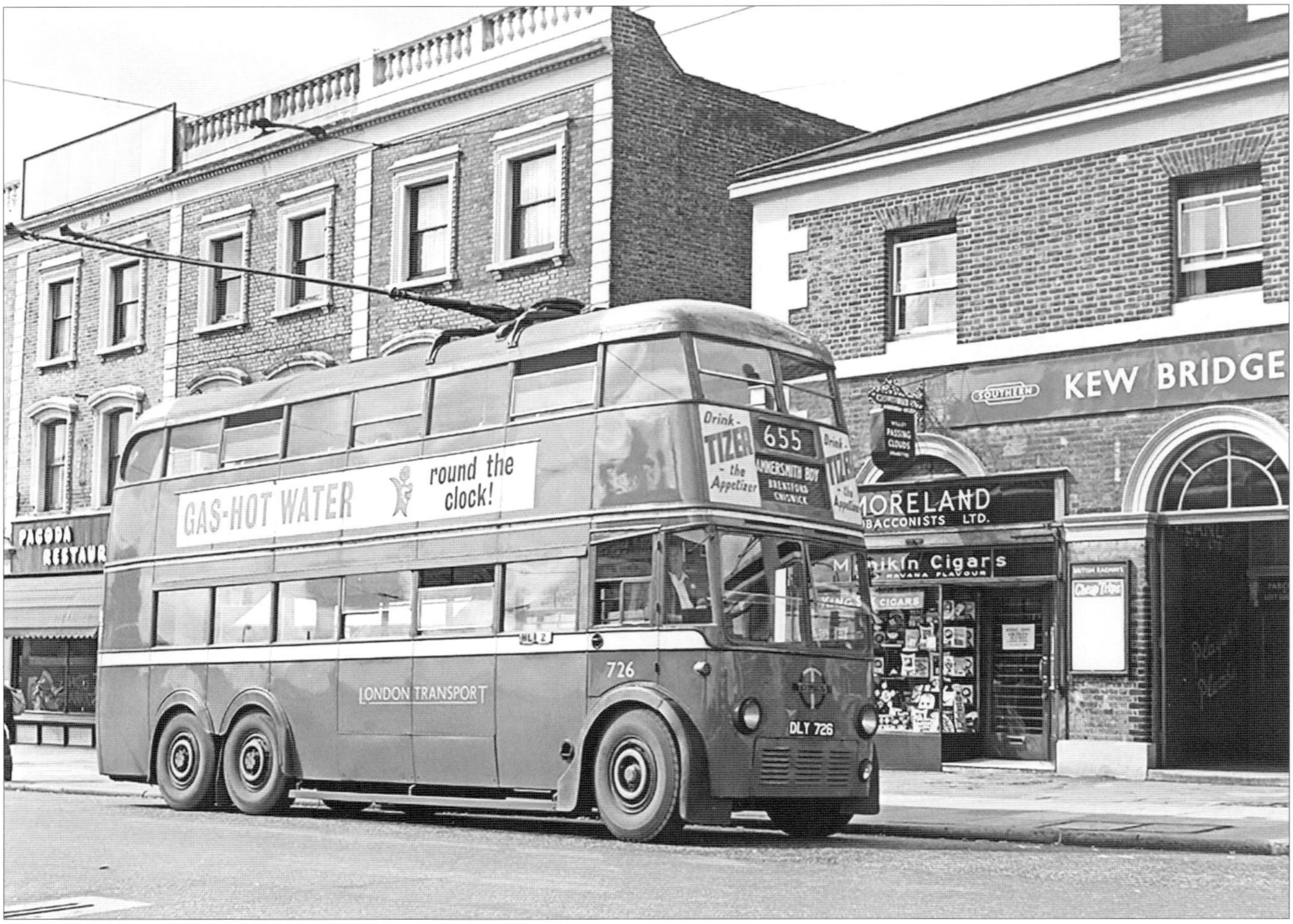
GAS-HOT WATER
round the clock!
Drink TIZER the Appetizer
655
HAMMERSMITH BDY
BRENTFORD
CHISWICK
LONDON TRANSPORT
726
DLY 726
KEW BRIDGE
SOUTHERN
MORELAND
TOBACCONISTS LTD.
Manikin Cigars
PAGODA RESTAUR

Opposite Top: Aldgate, 10th September 1958. L3 No. 1466 on the 569 is only going as far as its depot at Poplar, rather than North Woolwich. Behind is RTW 83, also heading due east, following a route which the 15 still does, if somewhat shortened. *(A. R. Gault)*

Opposite Bottom: Leyland F1 class trolleybus, MCW bodied Leyland, No.726 of 1938 at Kew Bridge on its way from Acton by way of Brentford to Hammersmith Broadway. Although there were many classes of trolleybus they were almost all built to a standard, most handsome design, on either a Leyland or AEC chassis, with little variation over the years. *(A. R. Gault)*

Above: Showing off its elegant lines Q1 No.1820 pauses in Teddington on its way to Twickenham with the background of what were then household names found all over the London suburbs, 3rd August, 1959. Burton Taylors presents an impressive, typical between the wars face, whilst it is still hard to believe the once universal Woolworths has vanished entirely, 3rd August, 1959. *(A. R. Gault)*

Above: Q1 No.1820 looks magnificent, posed beside the YMCA at Kingston on 3rd August, 1959 on its way to Tolworth. There is still a YMCA in the town nowadays moved some time ago to beside the river. two of the batch of HYM registered Q1s have been preserved *(A. R. Gault)*

Opposite Bottom: The very last trolleybuses to enter service in London were the second batch of Q1s, dating from 1952. Identical in every respect to their HYM registered, 70 seat predecessors, No.1880 is seen at Kew Bridge, working the 667. Allocated to Fulwell, Isleworth and Hanwell depots, these, too, were sold to Spain. *(A. R. Gault)*

Above: L3 No.1483 leaving the turning circle at Tolworth Red Lion into Ewell Road at Tolworth on the Kingston 603. No.1483, dating from late 1939, had been transferred into Fulwell depot to replace the sold Q1s in 1961 and would survive until the last day of the London trolleybus system, 8th May, 1962. *(A. R. Gault)*

Below: B1 90 is one of a batch of 30 short wheelbase trolleybuses designed for hilly routes. They worked on the 654 between Crystal Palace and Sutton out of Carshalton (CN) depot. The climb up Anerley Hill towards Crystal Palace was quite a challenge. *(A. R. Gault)*

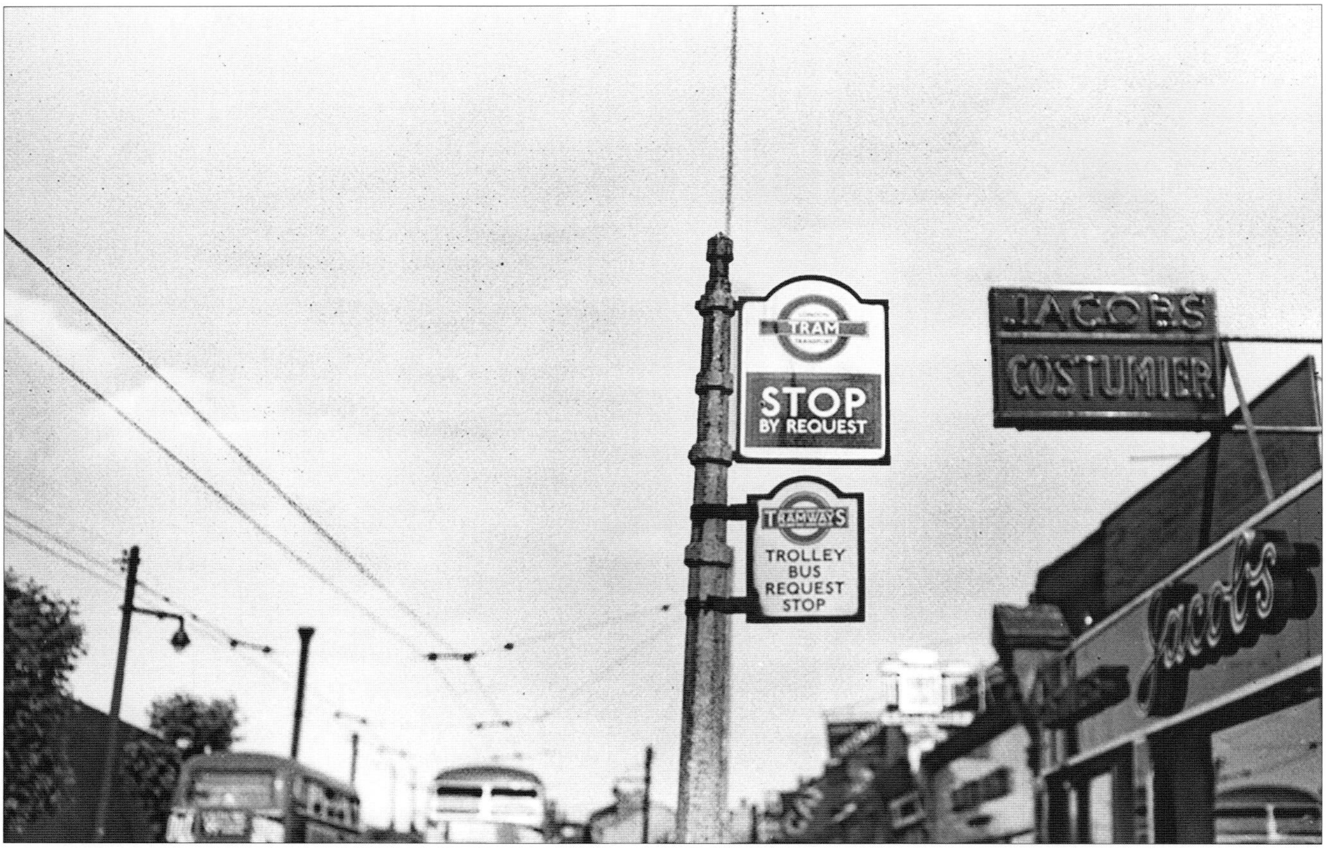
TRAM
STOP
BY REQUEST
TRAMWAYS
TROLLEY
BUS
REQUEST
STOP
JACOBS
COSTUMIER
Jacobs

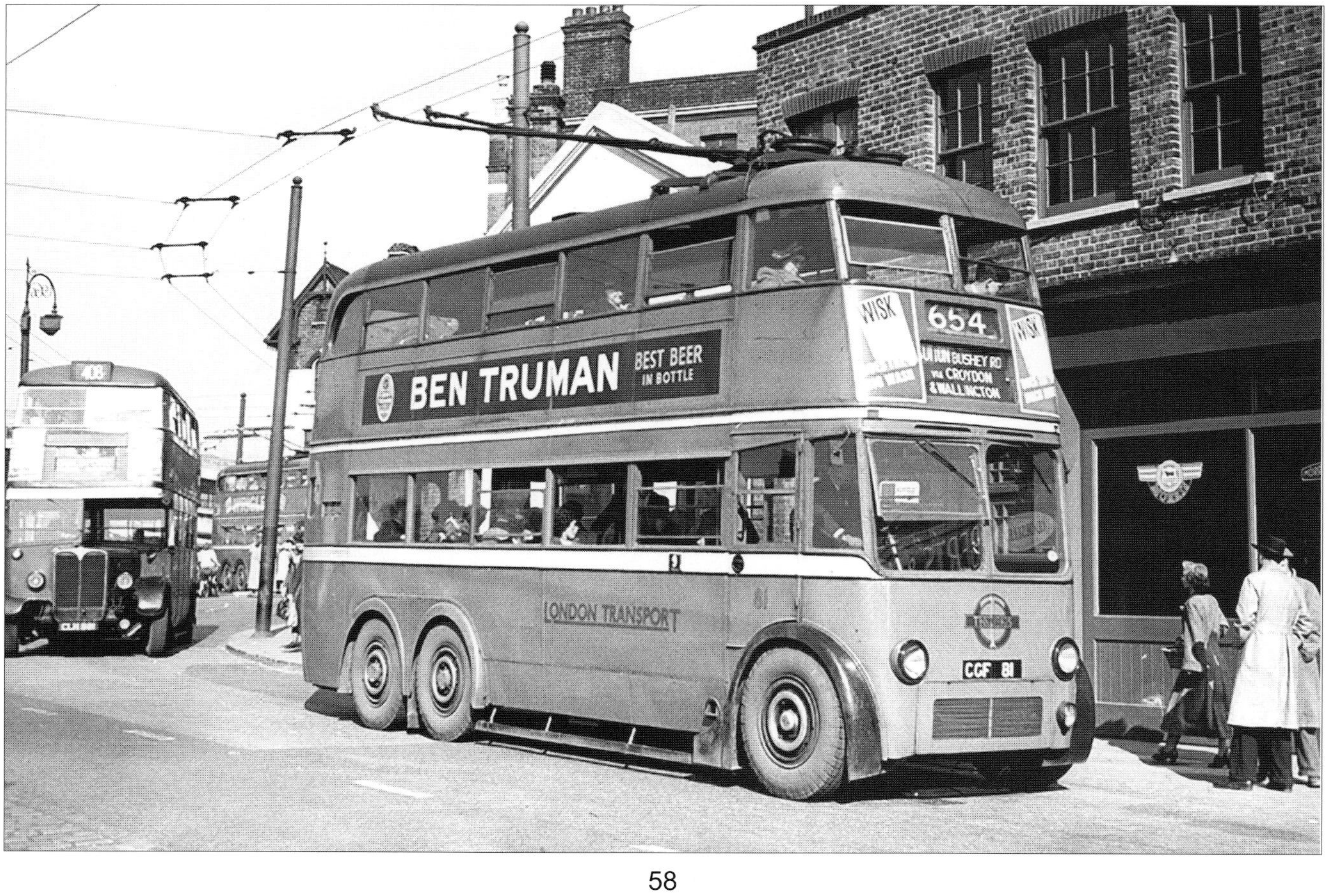
654
BUSHEY RD
via CROYDON
& WALLINGTON
BEN TRUMAN
BEST BEER
IN BOTTLE
LONDON TRANSPORT
CGF 81
408

Opposite Top: A most unusual stop, pre-London Transport and serving a trolleybus route even though it declares itself belonging to the tramways. We do not know the location but it could be in the Kingston area. *(A. R. Gault)*

Opposite Bottom: Tamworth Road, West Croydon in 1950. B1 trolleybus No.81 of Carshalton Depot has just crossed over the main London Road along which tram routes 16, 18 and 42 will continue to pass for another few months, followed by Country Area STL2190 of 1936, whilst another trolleybus, working route 630, almost certainly a 1936 vintage D2 from Hammersmith (HB), is following. The B1s, dating from the end of 1935, were short wheelbase Leylands, with special brakes designed specifically for tackling the steep Anerley Hill leading up to Crystal Palace. No.81 retains the old running number rather than the later depot/garage code adopted when the tram and trolleybus department was absorbed by the motor bus one in 1951. *(A. R. Gault)*

Above: No.455, an all-Leyland D2 of late 1936, based at Hammersmith (HB) depot, is seen at the West Croydon terminus of the 630 route. It had claim to be London's longest regular one, extending all the way through Mitcham, Tooting and Wandsworth, across the Thames at Putney, to Hammersmith, Shepherd's Bush and terminating at the most curiously named "Near Willesden Junction," a destination description which was replaced in its final years by the infinitely more prosaic Harlesden. *(A. R. Gault)*

Above: B1 No. 69 stands at the top of Anerley Hill having successfully achieved the summit and takes a breather before zooming back down towards South Norwood, Croydon and points west to Sutton. The area was known as Upper Sydenham before the wonderful glass and iron structure which had housed the Great Exhibition of 1851 in Hyde Park was moved there and gradually Sydenham assumed the name Crystal Palace. Trolleybuses served the Palace for just one year before a spectacular fire consumed the fabulous building. *(A. R. Gault)*

Opposite Top: Class H1B 790B seen here in Plumstead in May 1954 on route 696. Bexleyheath depot was home at one time to no less 29 rebodied trolleybuses damaged during the war, which added considerable interest to the vehicles on routes 696 and 698.

Entering service in 1938 H1 Class 790 was built with a MCCW body. However, 790 was one of a number of vehicles badly damaged when a Flying Bomb (V1)/'Doodlebug' struck Bexleyheath depot on 29th June 1944. 790 was subsequently rebodied by East Lancs and reclassified as H1B, now carrying the number 790B. It was with withdrawn from Bexleyheath in March 1959 when trolleybus operation ceased on March 3rd as part of the Stage 1 trolleybus to bus conversion programme. The 696 (Woolwich – Dartford) route was renumbered to 96 and operated by RTs. The route was converted to OPO in November 1971 with new DMSs. *(A. R. Gault)*

Opposite Bottom: Rebodied D2 No.402 passing the Harrow Inn, Abbey Wood, 19th May, 1954. This Leyland vehicle was another wartime casualty, its new body being supplied by Northern Coach Builders and delivered at the very end of 1945; London Transport had to fit the seats. *(A. R. Gault)*

Above: This picture depicts Trolleybus K1 1109 on route 683, at the Stamford Hill terminus ready to return to Moorgate. It only ran on weekdays and after 1956 it just ran at peak hours. It was less frequent than the other routes through Stoke Newington with a roughly 15 minutes headway. It was also the route to turn right at Dalston Junction into Balls Pond Road. The 683 was never a paying proposition and it was one of the many cuts following the strike of May 1958. The reason it was not a success was that it was paralleled by the more frequent 76 bus route, and when it came off in January 1959 it was not replaced. *(A. R. Gault)*

Opposite Top: We next encounter E2 No.624, having reached Stratford on 19th May, 1954, taking a breather before returning to North Woolwich. It is of passing interest to note that the Queen used Cerebos salt. Also that adverts were never applied to the rear upper deck of London trolleybuses on account of the two little knobs protruding either side of the indicators designed to prevent damage should the rear emergency window need to be opened. *(A. R. Gault)*

Opposite Bottom: 43 trolleybuses intended for South Africa never reached there on account of the voyage being considered low priority in wartime and instead helped fill the gap caused by war damage to the London fleet. Even repainted in London livery there was no mistaking them as being very non standard, not least their 8ft width. The Metropolitan Police, somewhat reluctantly, gave them dispensation, but only to operate far from central London on two routes in the Barking area. Not all the South Africans were identical, they being originally intended for different cities and fleets. The somewhat more curvaceous No.1753 is waiting over at the Barking terminus of the 693 before setting off on the 25 minute run by way of Ilford High Road to Chadwell Heath. *(A. R. Gault)*

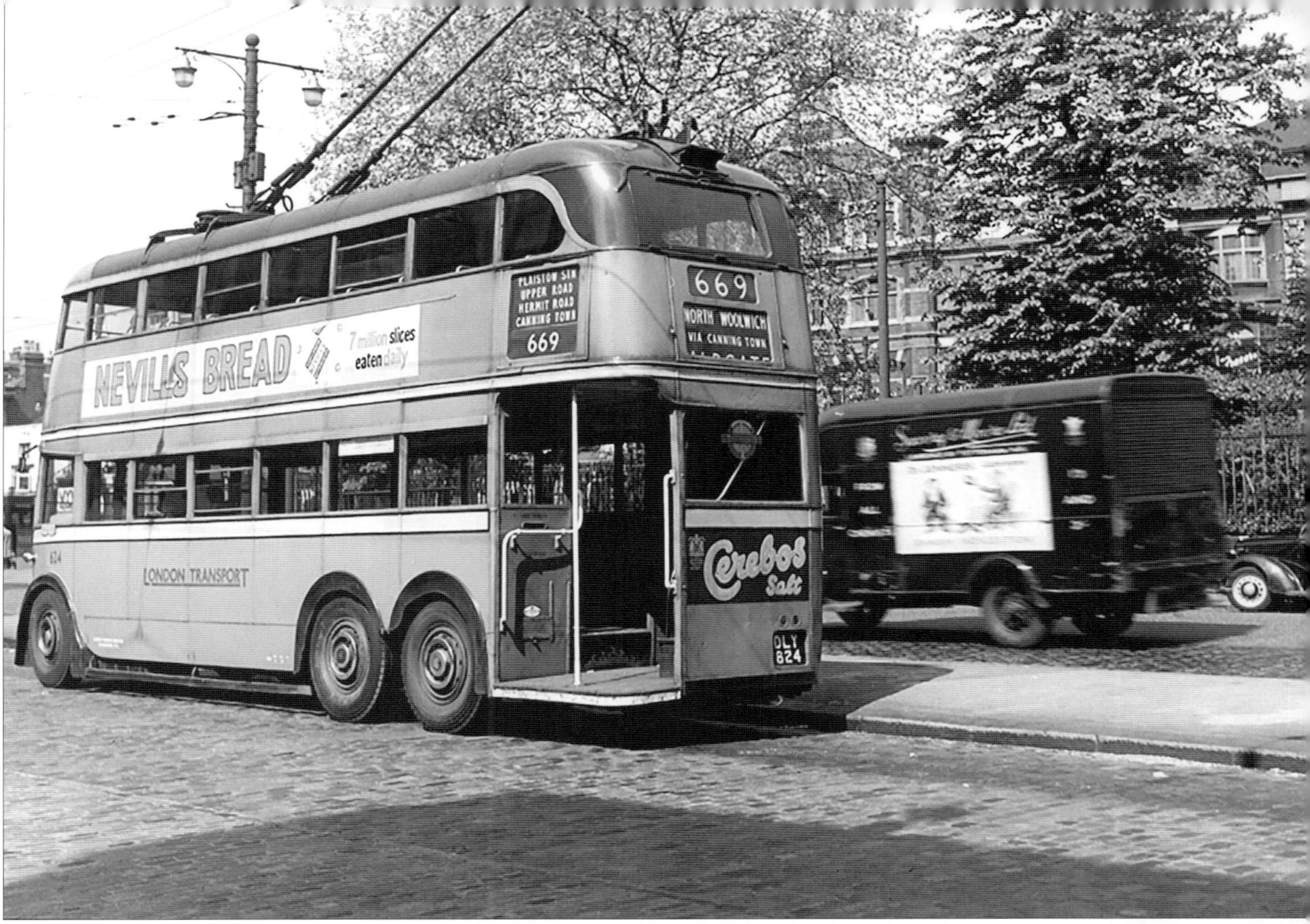
PLAISTOW STN
UPPER ROAD
HERMIT ROAD
CANNING TOWN
669
669
NORTH WOOLWICH
VIA CANNING TOWN
NEVILLS BREAD
7 million slices
eaten daily
LONDON TRANSPORT
Cerebos
Salt
DLY
824

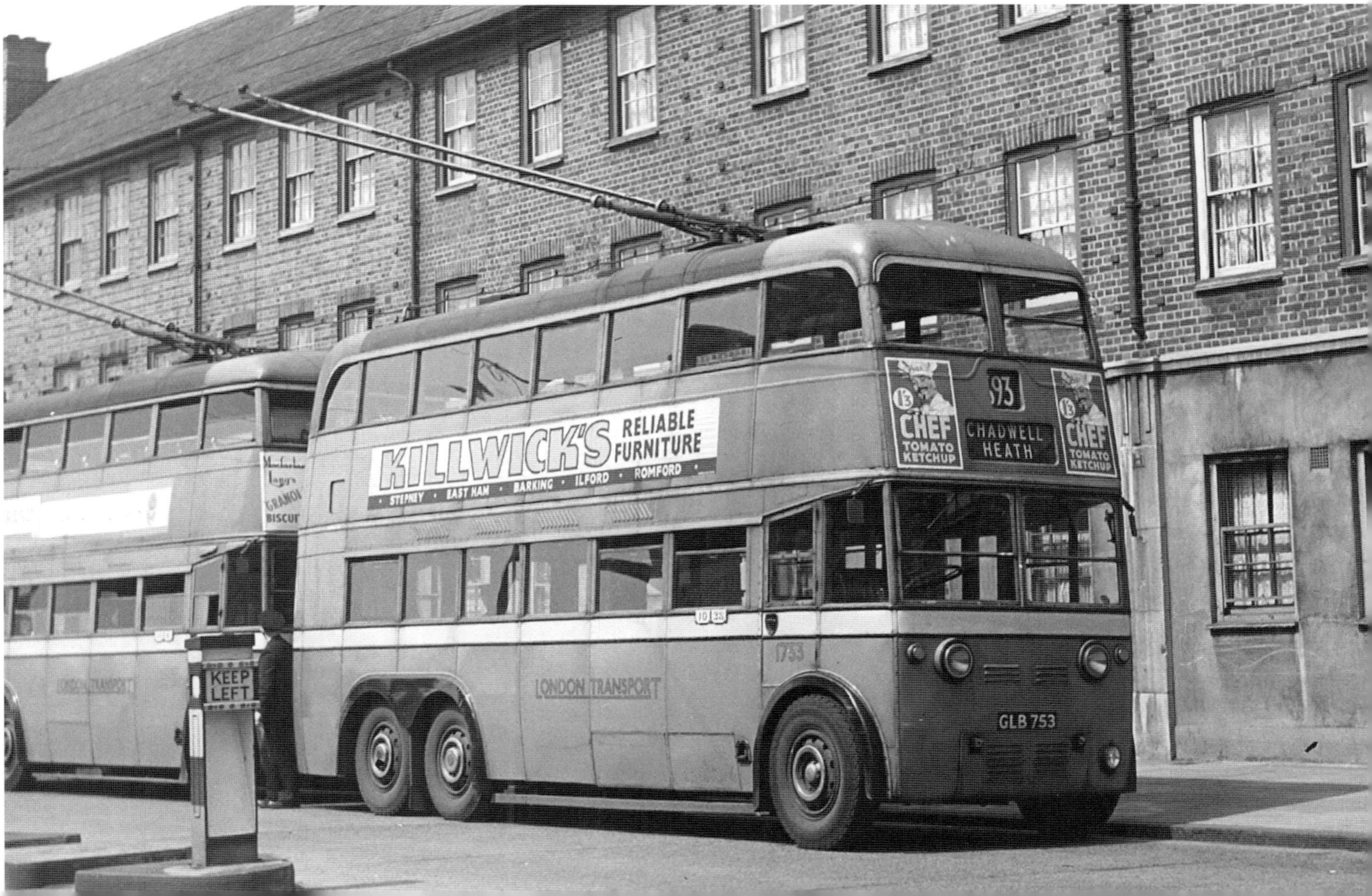
KILLWICK'S
RELIABLE
FURNITURE
STEPNEY · EAST HAM · BARKING · ILFORD · ROMFORD
CHEF
TOMATO
KETCHUP
CHADWELL
HEATH
CHEF
TOMATO
KETCHUP
1753
LONDON TRANSPORT
GLB 753
KEEP
LEFT
LONDON TRANSPORT

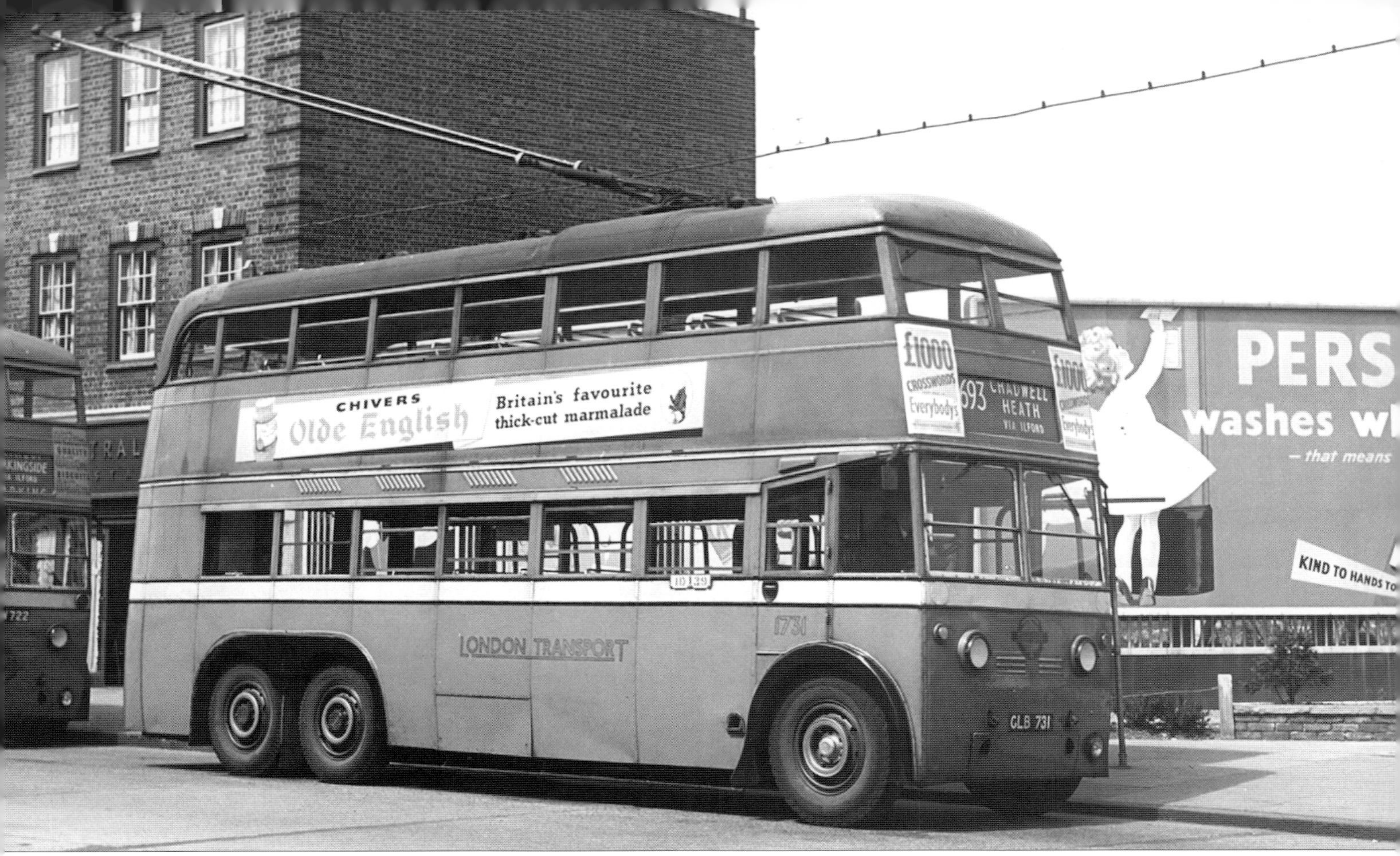
CHIVERS
Olde English
Britain's favourite
thick-cut marmalade
£1000
CROSSWORDS
Everybodys
693 CHADWELL HEATH
VIA ILFORD
PERS
washes wh
– that means
KIND TO HANDS TO
LONDON TRANSPORT
1731
GLB 731

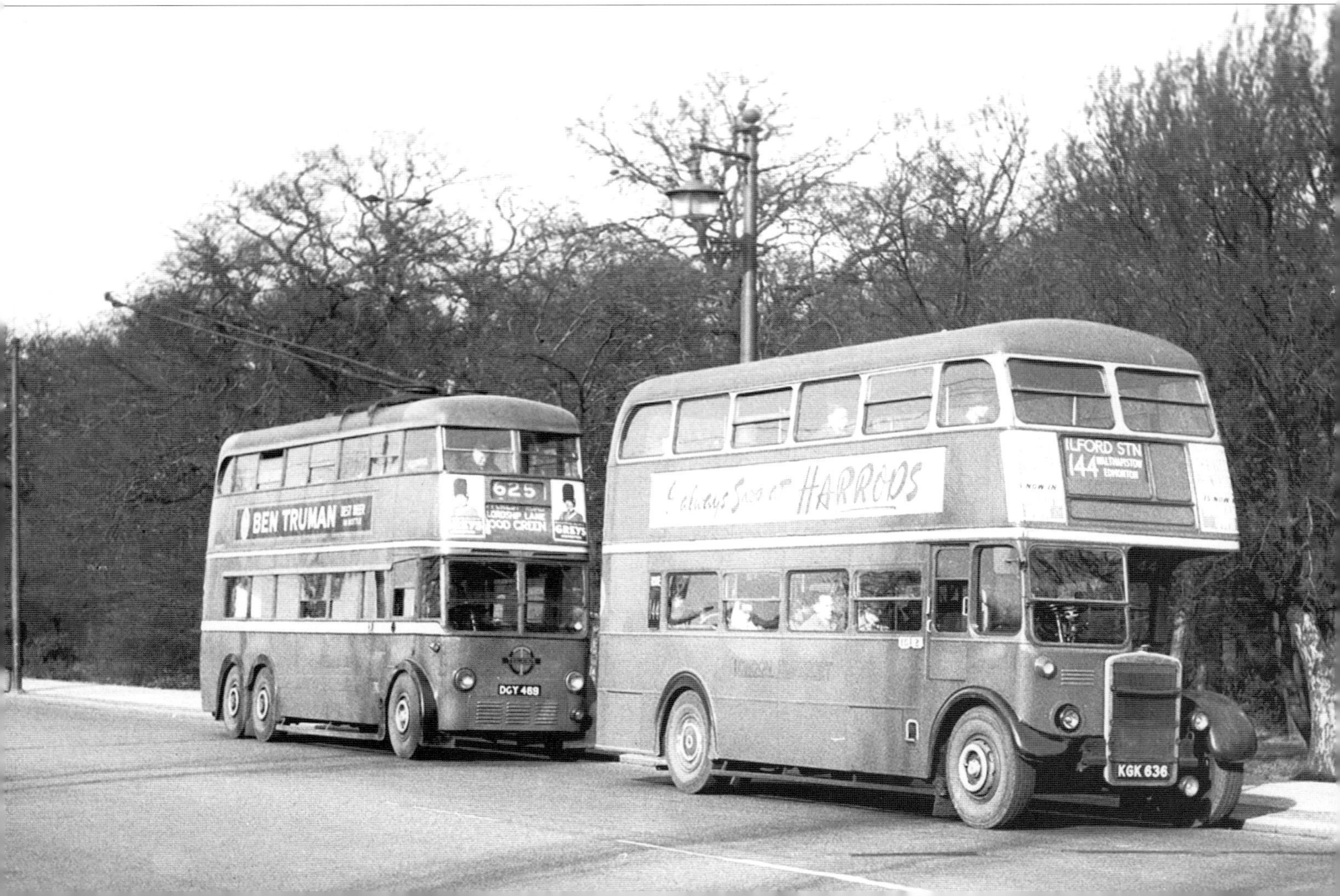
625
LORDSHIP LANE
BEN TRUMAN
GREYS
HARRODS
ILFORD STN
144
DGY 469
KGK 636

Opposite Top: The more upright No. 1731 is also bound for Chadwell Heath. London probably went in for more adverts plastered all over its vehicles than other transport concerns, bringing in a huge income and adding splashes of colour and, sometimes, striking designs, but none could compete with the Persil one on the hoarding behind, which was a classic, a perfect mix of words and visuals. *(A. R. Gault)*

Opposite Bottom: There were relatively few places where London trolleybuses terminated beside a forest but D2 469, along with RTW136, demonstrates much the best known. This c1950 photograph was taken at Woodford on the edge of Epping Forest. *(A. R. Gault)*

Above: No.1321, a February, 1939 vintage K2, trundles across the cobbles at Hackney on its way to the docks, 12th April, 1958. *(A. R. Gault)*

Above: Clapton 25th July, 1958. Leading the procession is an all-Leyland K1 No.1291 of Clapton depot (trams and trolleys lived in depots, buses in garages, their crews quite often in council houses, seldom in stately homes), dating from 1939, bound for Bloomsbury, very close to the northern entrance of the Kingsway Subway, and as near to the West End as trolleybuses were allowed, followed by a Leyton (T) RT and RTL 765. *(A. R. Gault)*

Opposite Top: No.1255 on a short working of the 677 from West India Docks to Smithfield, 25th July, 1958. In the 1950s winning on the football pools was the favoured way of finding yourself owner of a fortune without having to work for it. *(A. R. Gault)*

Opposite Bottom: The 653, which weaved its way from Aldgate around a large slice of north London and terminated 64 minutes later at Tottenham Court Road, was one of the busiest of London's trolleybus routes requiring 49 vehicles, Monday to Friday. MCW bodied J2 AEC, No.1000 with a unique registration, dating from 1938, was based at Highgate, which was home to more trolleybuses than any other depot, beating West Ham by just one. RF332, behind, also possessed an MCW body. *(A. R. Gault)*

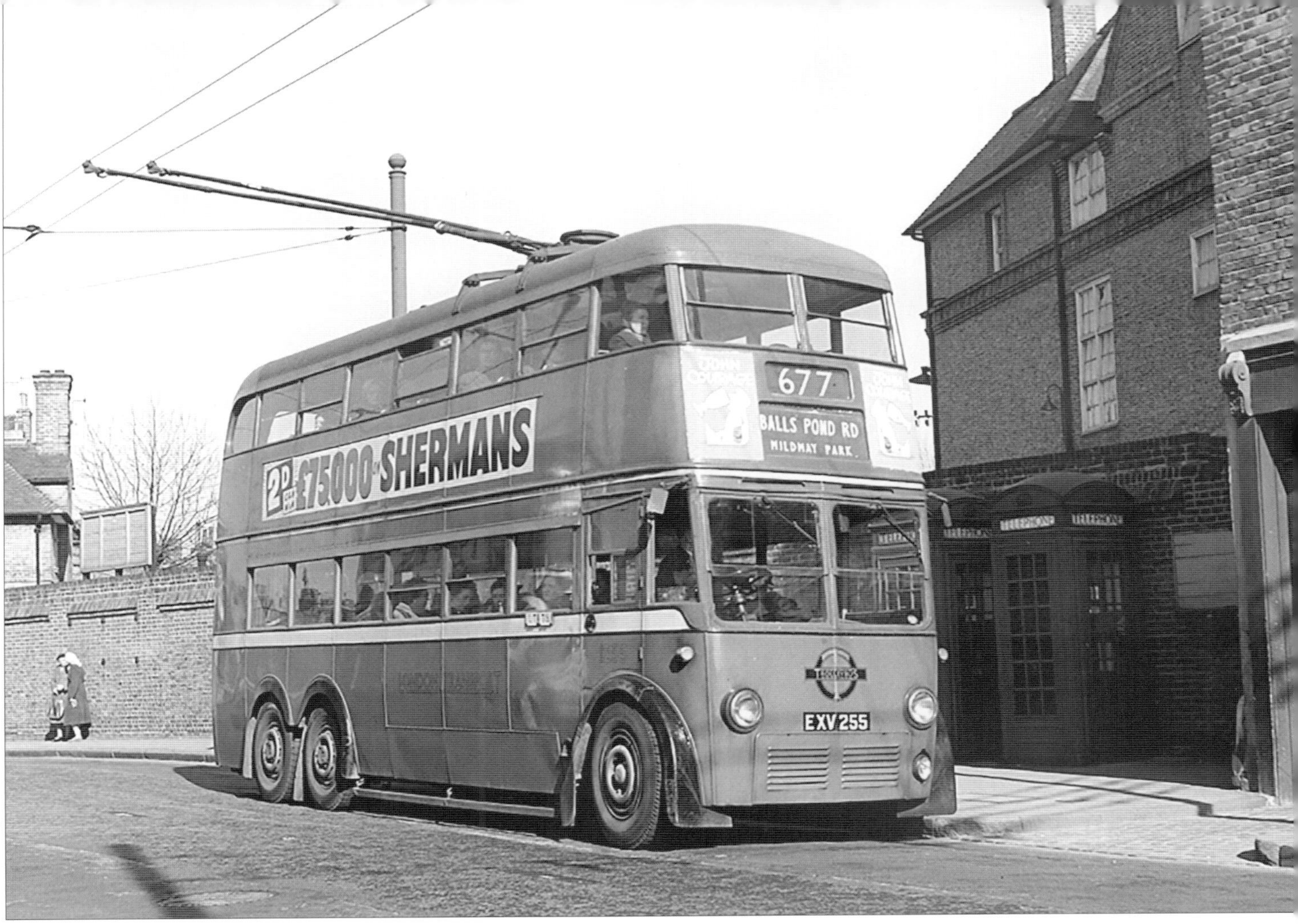
677
BALLS POND RD
MILDMAY PARK
SHERMANS
EXV 255
TELEPHONE

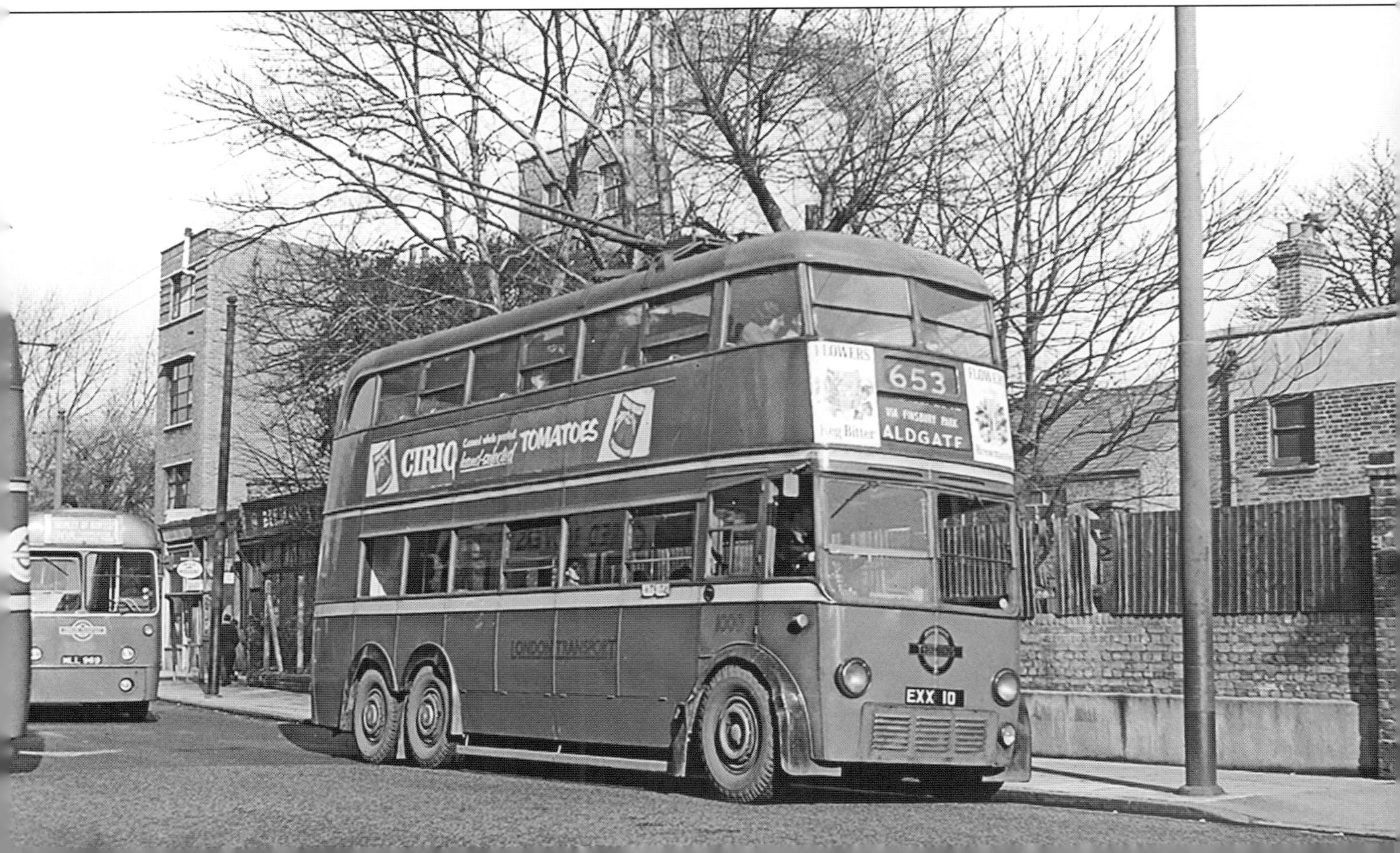
653
VIA FINSBURY PARK
ALDGATE
FLOWERS
Keg Bitter
CIRIQ
TOMATOES
LONDON TRANSPORT
EXX 10

Above: The entire complement of passengers and both driver and conductor appeared to have abandoned BRCW bodied AEC N1 No.1578 on 24th July, 1958 for the Kings Arms, situated in Bow Road. Has the occupant of the grand baby carriage joined them as well, one wonders? The N1s, dating from 1939/40, moved westwards and ended their service at either Colindale or Stonebridge depots in January, 1962. *(A. R. Gault)*

Opposite Top: Bow N1 No.1570 makes a left turn beside the Westminster Bank into Fairfield Road, Bow on a bright, sunny July day in 1958. Route 695 would be withdrawn in January, 1959, being a weekdays only route which could be covered by other routes, six months before these too would go, replaced by Routemasters. *(A. R. Gault)*

Opposite Bottom: A scene in the Lower Clapton Road, E5 on 12th April, 1958 with two K2s, Nos. 1351 and 1175. One wonders how many trolleybus passengers really did shop at Harrods. *(A. R. Gault)*

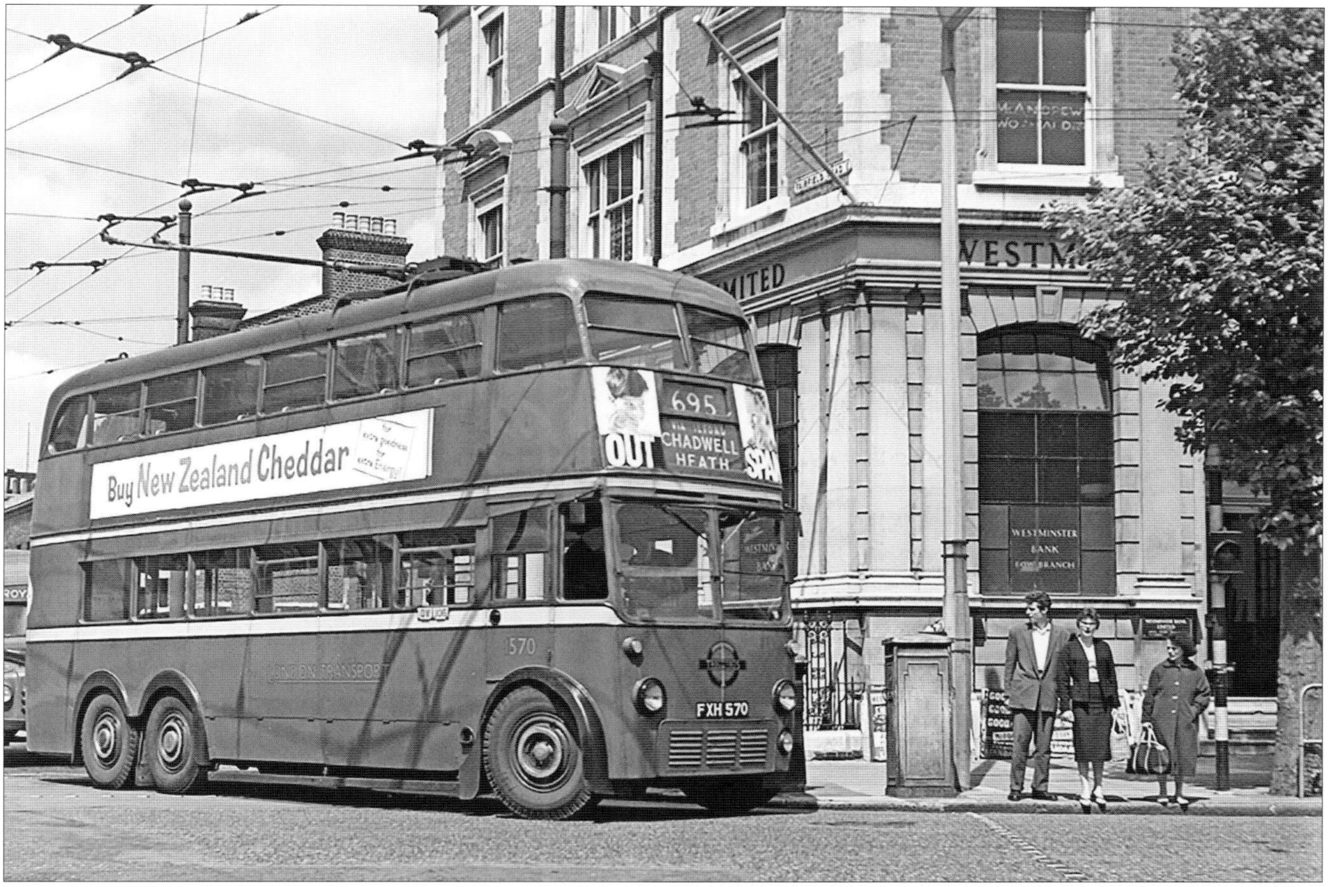
Buy New Zealand Cheddar
695
OUT
CHADWELL
HEATH
570
FXH 570
WESTMINSTER
BANK

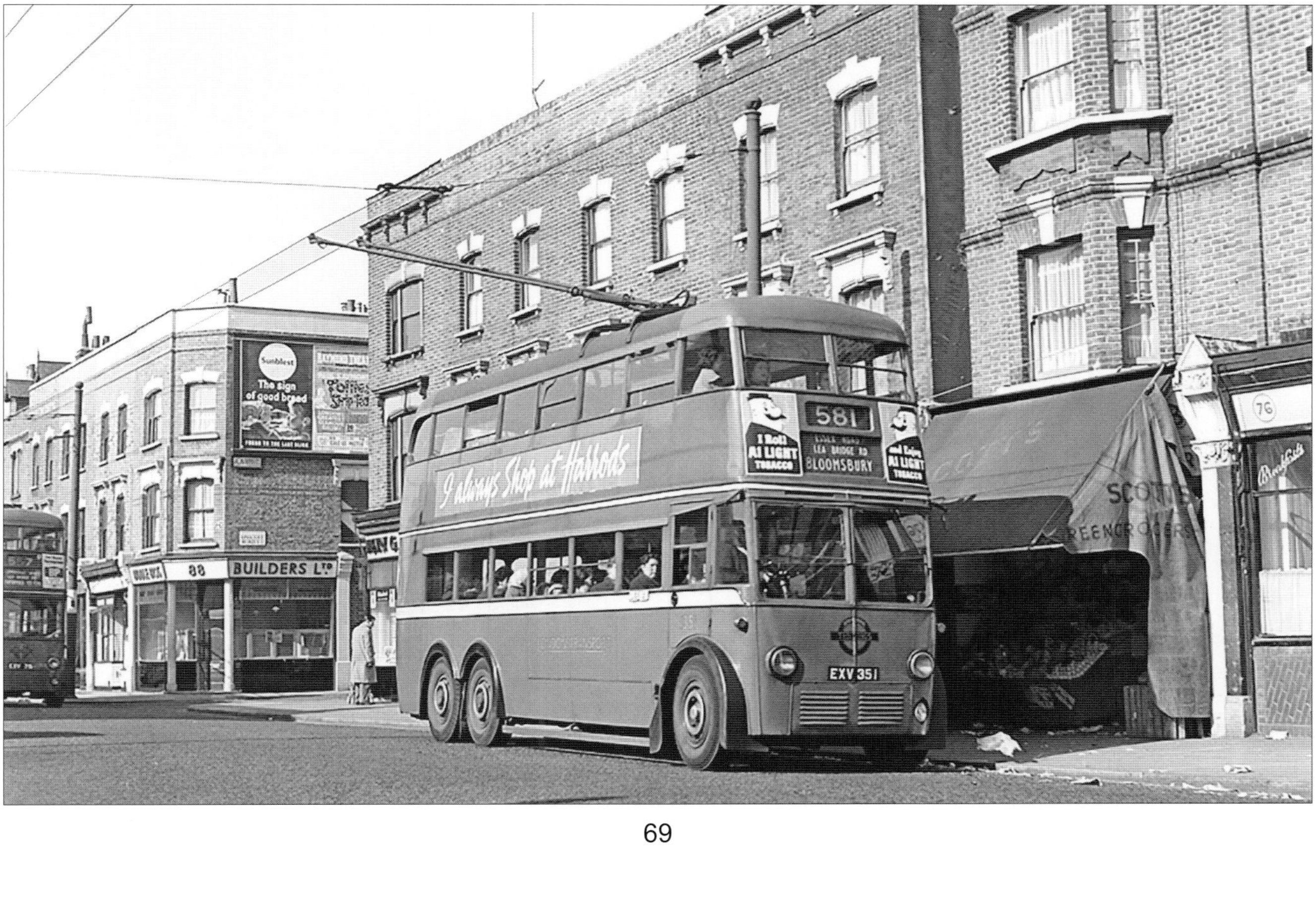
Sunblest
The sign of good bread
BUILDERS Ltd
88
581
LEA BRIDGE RD
BLOOMSBURY
AI LIGHT
TOBACCO
I always Shop at Harrods
EXV 351
76

A large fleet of service vehicles has always been essential to keep the show on the road, but particularly in tram and trolleybus days. An AEC Mercury wire maintenance vehicle (tower wagon) of 1935 at Manor House, 24th July, 1958. *(A. R. Gault)*

Highgate depot's rather careworn MCW bodied AEC C1 class No.169, of 1935, stands at the Parliament Hill Fields terminus of the 615, 25th April, 1953. In 30 minutes it will arrive at its city terminus. Service cuts in 1955 ensured that the last of the C1 class, already scheduled for withdrawal, were quickly condemned, slightly more modern vehicles taking their place. It is advertising two very popular, if contrasting, contemporary thirst quenchers. *(A. R. Gault)*

Within the 569's 41 minute run there were at least two short workings, to Poplar depot and to Silvertown station, the penultimate Eastern Region one before North Woolwich, railway and trolleybus route running parallel for much of the final section of their journeys. Trolleybus No.1514 seems to be of two minds, rear and front indicators being in disagreement. Trams and then trolleybuses had to terminate at Aldgate on account of the City's rather fierce objection to 'working class' forms of transportation, whilst the motor bus was able to continue on right through the City streets and on to the West End. *(A. R. Gault)*

Across the water at North Woolwich, 19 May, 1954, we find two West Ham based AEC trolleybuses, E2 No.624 and E1 No.569, one working the 569 to Aldgate, the other the 669 to Stratford. Behind, beyond the 1939 vintage big Austin, is the entrance to the foot tunnel under the River Thames. *(A. R. Gault)*

Above: Highgate's all-Leyland D3 No.544 of 1937, working the 639, pauses alongside the City Arms in City Road on 24th September, 1958 on its way to Moorgate. 15th April, 1959 saw the end of the D3 class, replaced by more modern vehicles; the 639 lasted until 1 February, 1961. The handsome block of buildings incorporating the City Arms is still there although the City Arms itself is now a Co-Op. *(A. R. Gault)*

Opposite Top: Trolleybus Class L2 1370 is seen passing the offices and showroom of Metalferm Ltd on Grays Inn Road on route 513 heading towards Holborn Circus on 6th October 1958. The twin routes 513 and 613 ran between Hampstead Heath (South End Green) and Parliament Hill Fields. The two termini were barely a kilometer apart, separated by Parliament Hill Fields, the southern extension of Hampstead Heath. The only difference between the two routes was that that 513 ran from Hampstead Heath to Holborn Circus via Kings Cross and Grays Inn Road. From Holborn Circus it continued on back to Kings Cross via Faringdon Road and Kings Cross Road, thus forming a loop. From Kings Cross the routes ran parallel as far as Kentish Town where they divided; the 613, on the other hand ran in the opposite direction reaching Holborn via Faringdon Road before heading north west back to Hampstead Heath. Outside the peak hours they were two relatively lightly used routes. *(A. R. Gault)*

The changeover from tram to trolleybus traction took place on Sunday 10th July 1938. Prior to this, the two tram routes had been the 3 and 7, which corresponded to the 513 and 613 respectively. The only difference was that both tram routes ran via Grays Inn Road and terminated at Holborn at the end of Grays Inn Road by Chancery Lane Underground station. The trolleybuses would not have been able to terminate as there was nowhere to turn, so a loop was introduced via Faringdon and Kings Cross Roads.

Parked in front of 1370 on Grays Inn Road is a Ford Popular car and behind it is an Austin A30. *(A. R. Gault)*

Opposite Bottom: No. 306, a BRCW bodied AEC of 1936 of Stonebridge Depot, pauses outside the Desborough Arms in Harrow Road on its way from Edgware to Paddington Green on 4th October, 1958. The 664 ended on 6th January, 1959 resulting in the withdrawal of most of the remaining C3s. The curious posters on either side of the destination indicator were simply an LT design filling an empty space. *(A. R. Gault)*

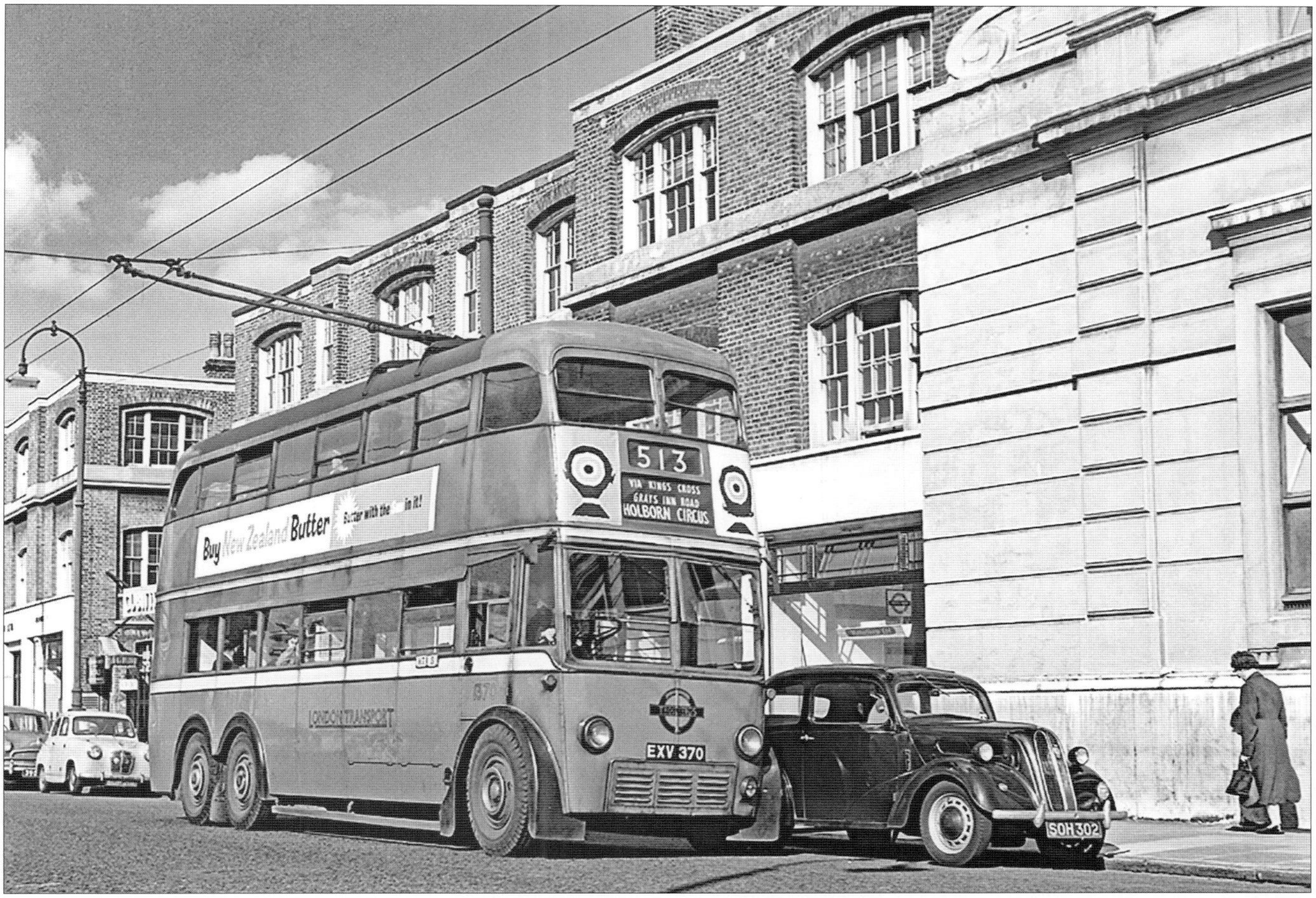
513
VIA KINGS CROSS
GRAYS INN ROAD
HOLBORN CIRCUS
Buy New Zealand Butter
LONDON TRANSPORT
EXV 370
SOH 302

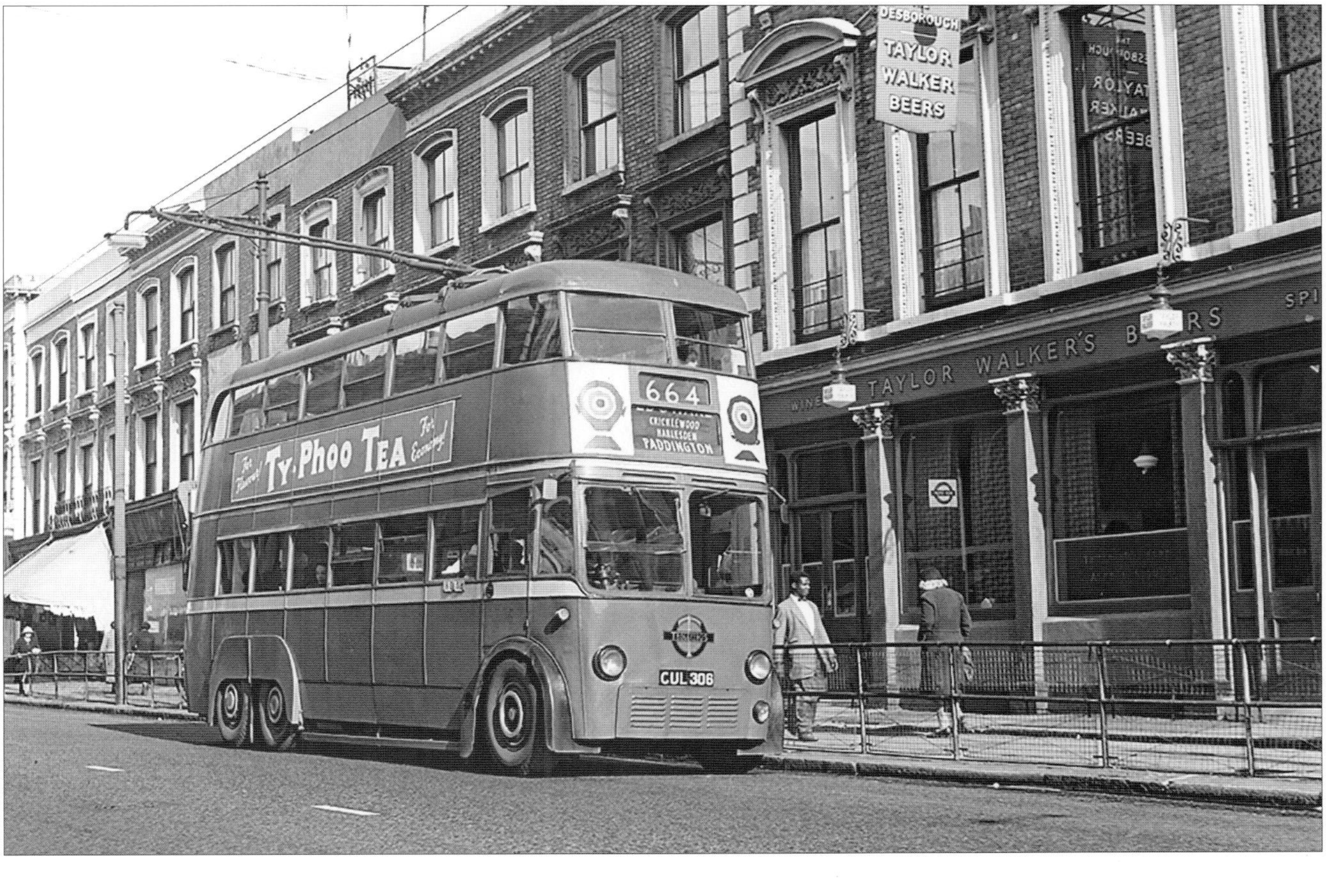
DESBOROUGH
TAYLOR
WALKER
BEERS
664
CRICKLEWOOD
HARLESDEN
PADDINGTON
For Flavour!
Ty-Phoo TEA
For Economy!
TAYLOR WALKER'S BEERS
CUL 306

On a bright, sunny afternoon Highgate K1 No.1139 of October, 1938, finds itself on the same cobbles in the Caledonian Road. It is perhaps fitting, giving where it is standing, that the posters either side of the front indicator are adverts for British Railways. *(A. R. Gault)*

Above: Walthamstow (WW) C3 No. 330, an MCW bodied AEC of October, 1935, at the Wood Green terminus of route 625, 8th April, 1953, sporting a very local advertisement. *(A. R. Gault)*

Below: Finchley all-Leyland J1 type No.919, delivered in March, 1937, in the Caledonian Road on the bridge spanning the tracks leading to King's Cross station, 2nd August, 1958. The finials of the signals can just be seen poking over the bridge parapet. It was still possible to detect where tram tracks had been set between the cobbles. *(A. R. Gault)*

Above: K2 1222 is seen in Stoke Newington High Street on Sunday route 649A, operating from Wood Green to Liverpool Street to serve the Sunday markets on Petticoat Lane, Brick Lane and Sclater Streets. Behind the trolleybus is the Vogue cinema, one of no less than 11 cinemas between Stamford Hill and Dalston, a distance of about 4 kilometres. S.Apatoff is one of the many Jewish businesses in the area. During the week routes from Wood Green passing through Stoke Newington were 543 and 643 to Holborn Circus. *(A. R. Gault)*

Opposite Top: Fulwell's L3 1433 on route 601 Tolworth to Twickenham via Kingston. In the late 1950s the 8'foot wide trolleybuses which had run Fulwell's routes since 1948 were withdrawn and sold off to Spain. They were replaced by older vehicles such as L3 class. *(A. R. Gault)*

Opposite Bottom: Three for the price of one on Seven Sisters Road, Finsbury Park on 2nd August 1958. Leading is H1 No.761, followed by K2 No.1321. Keeping an eye on the comings and goings at the City Pram emporium is a Doctor Who type police box. *(A. R. Gault)*

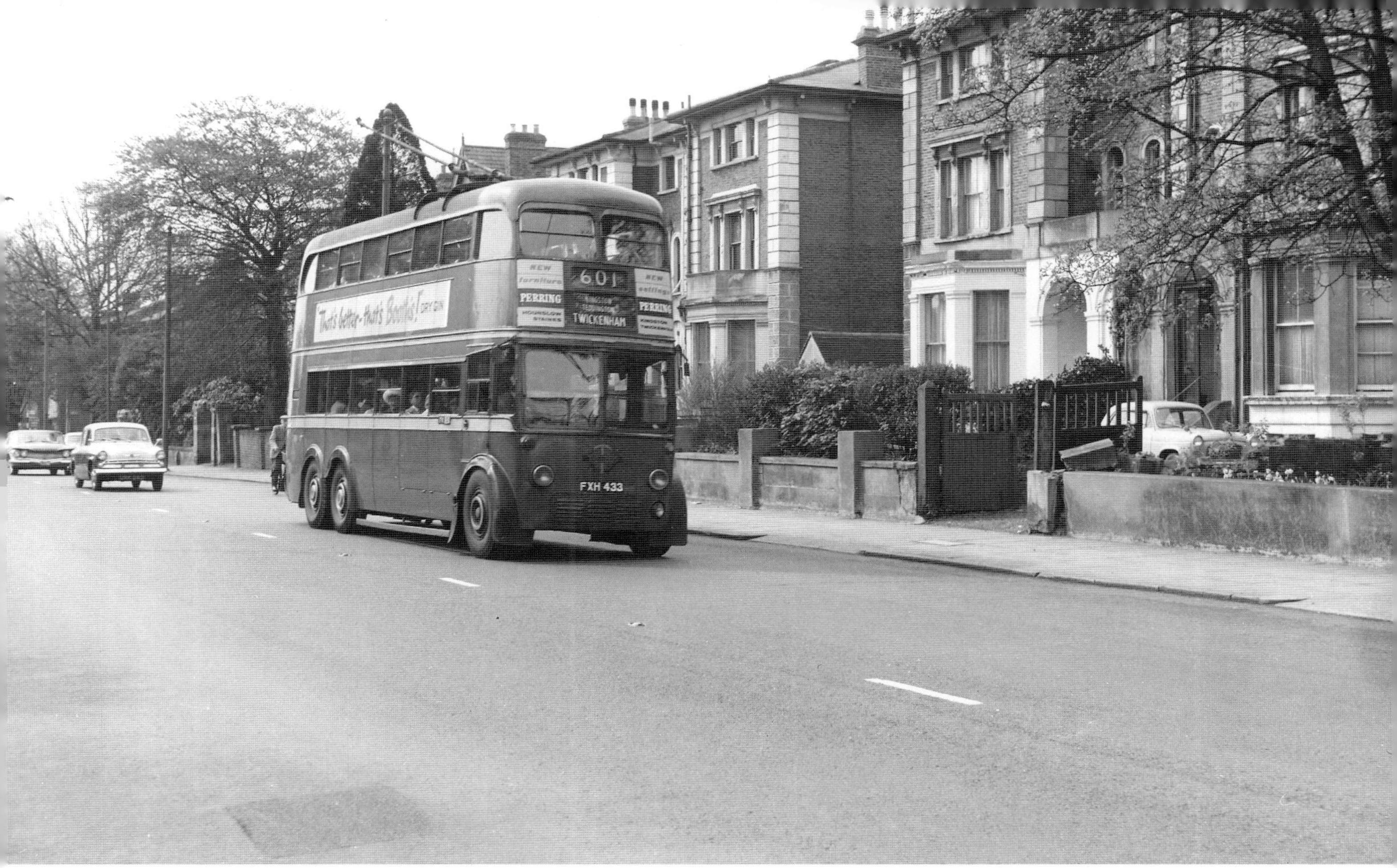

601
TWICKENHAM
PERRING
HOUNSLOW
STAINES
PERRING
"That's better - that's Booth's!" DRY GIN
FXH 433

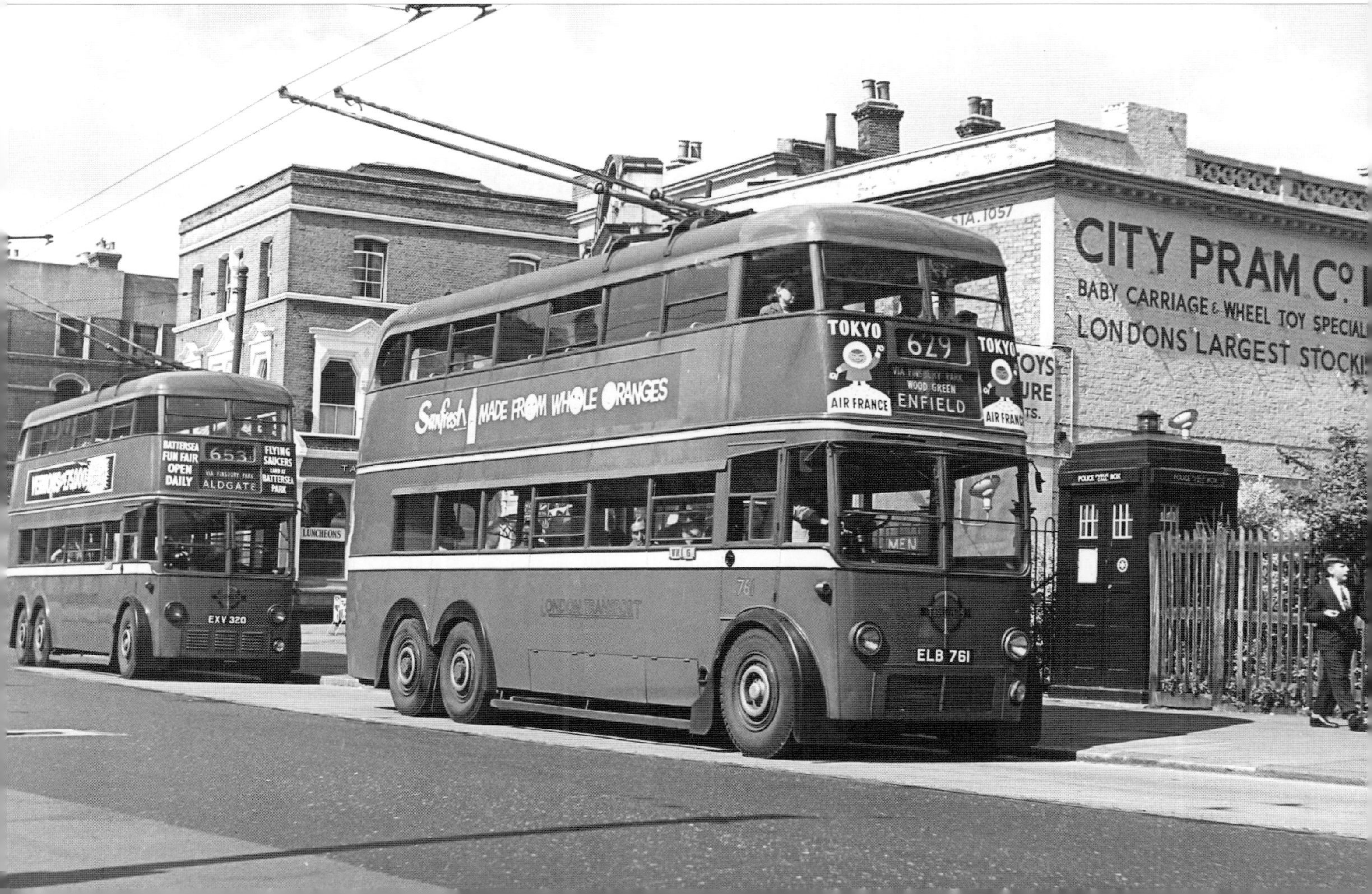

CITY PRAM Co
BABY CARRIAGE & WHEEL TOY
LONDONS LARGEST
TOKYO
629
VIA FINSBURY PARK
WOOD GREEN
ENFIELD
TOKYO
AIR FRANCE
AIR FRANCE
Sunfresh MADE FROM WHOLE ORANGES
LONDON TRANSPORT
ELB 761
653
BATTERSEA FUN FAIR OPEN DAILY
FLYING SAUCERS LAND AT BATTERSEA PARK
ALDGATE
EXV 320
LUNCHEONS
POLICE PUBLIC CALL BOX

Above: Queueing in the Seven Sisters Road, Manor House, 21st December, 1957. Although much in the area has changed, the rather impressive building forming a backcloth to C3 No.344, of Walthamstow depot (WW) which began service in August, 1936, still covering a Virginia Creeper route 623 to Woodford. No.344 was withdrawn in 1959. *(A. R. Gault)*

Opposite Top: Two K2s beside Stamford Hill depot, 9th July, 1955. It must be crew changeover time, the conductor of the rear, No.1208, vehicle is in conversation with its driver. No.1218 is leading. *(A. R. Gault)*

Opposite Bottom: C3 No.334 of Colindale depot (CE) in Harrow Road on 4th October, 1958 on its way to Paddington. Unlike a tram which, at the terminus, could just draw up in the middle of the road, the driver and conductor would exchange ends, and off it would go back whence it had come, a trolleybus had to find a convenient circle to circumnavigate. At Paddington the 664s were able to manoeuvre around the Sarah Siddons statue in the centre of Paddington Green. *(A. R. Gault)*

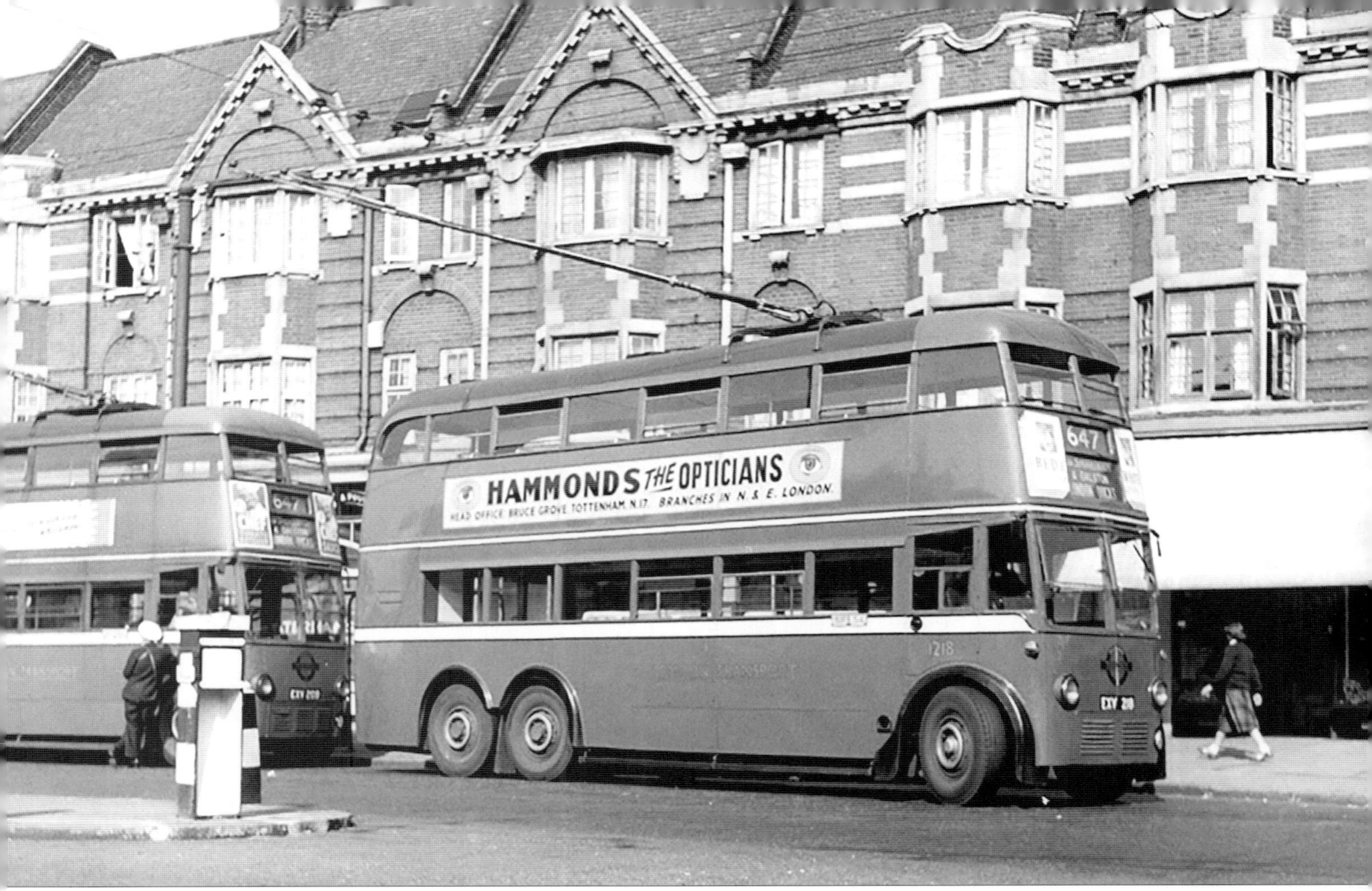
HAMMONDS THE OPTICIANS
HEAD OFFICE: BRUCE GROVE, TOTTENHAM, N.17. BRANCHES IN N. & E. LONDON.
647
1218
EXY 218

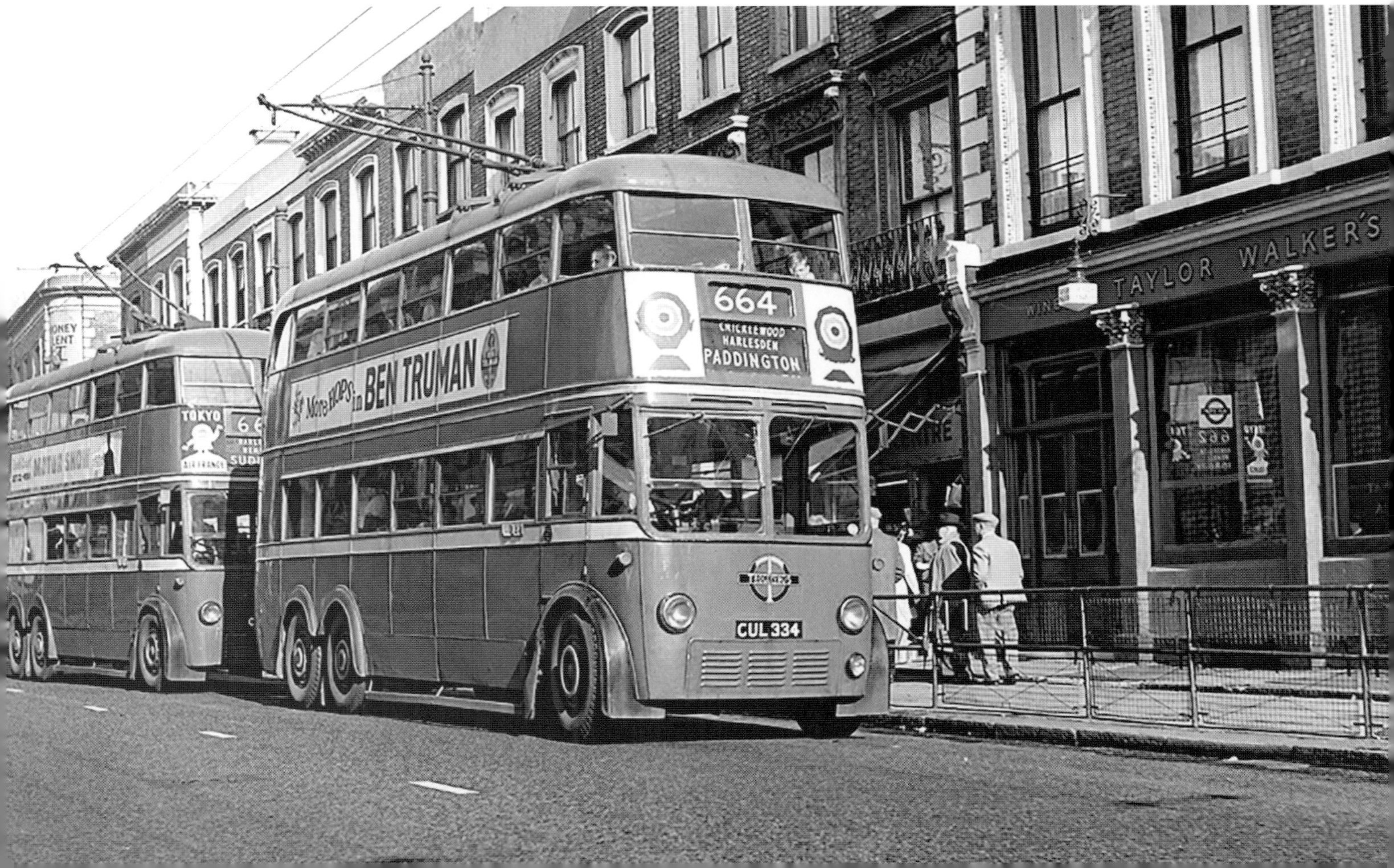
664
CRICKLEWOOD
HARLESDEN
PADDINGTON
BEN TRUMAN
TAYLOR WALKER'S
TOKYO
CUL 334

Aldgate, 10th September, 1958. Poplar based (PR) chassis-less MCW bodied AEC L3 bound for home. It could be argued that the standard trolleybus rear was aesthetically more pleasing than that of the RT which was more upright. The greater length of the former meant that even though it accommodated 70 seats, it had no need for so vertical a rear compared to the shorter, 56 seat motor bus. All a matter of opinion, I guess. *(A. R. Gault)*